revise
STANDARD GRADE
Physics

Jim Breithaupt, Brian Coleman
and Norman Stenton

with Tony Buzan

Hodder & Stoughton

A MEMBER OF THE HODDER HEADLINE GROUP

Key to symbols

As you read through this book you will notice the following symbols. They will help you find your way around the book more quickly.

 shows a handy hint to help you remember something

 shows you a short list of key facts

 gives worked examples to help you with calculations and equations

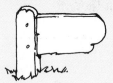 points you to other parts of the book where related topics are explained

| indicates credit level material

Standard Grade Physics and this Revision Guide

This revision guide is not intended to replace your textbooks. As tests and examinations approach, however, many students feel the need to revise from something a good deal shorter than their usual textbook. This revision guide is intended to fill that need. It covers all the Standard Grade Physics syllabus.

To use this book most effectively, make sure you read Tony Buzan's notes on 'Revision made easy'. Make a timetable for revision, using the contents list on page 3 to ensure you cover all the topics in the syllabus. Planned use of time and concentrated study will give you time for other activities and interests as well as work.

When the exam arrives, you should have confidence if you have revised thoroughly. In the examination room, attempt all the questions you are supposed to answer and make sure you turn over every page. Many marks have been lost in exams as a result of turning over two pages at once. If you suffer a panic attack, breathe deeply and slowly to get lots of oxygen into your system and clear your thoughts. Above all, keep your examination in perspective – it is important but not a matter of life or death!

We wish you success,

Jim Breithaupt
Brian Coleman
Norman Stenton

ISBN 0 340 77149 6

First published 2000
Impression number 10 9 8 7 6 5 4 3 2 1
Year   2006 2005 2004 2003 2002 2001 2000

The 'Teach Yourself' name and logo are registered trade marks of Hodder & Stoughton Ltd.

Copyright © 2000 Jim Breithaupt, Brian Coleman and Norman Stenton
Introduction ('Revision made easy') copyright © 2000 Tony Buzan

Learning outcomes in previews reproduced by kind permission of the Scottish Qualifications Authority.

All rights reserved. No part of this publication may be reproduced or transmitted in any form or by any means, electronic or mechanical, including photocopy, recording, or any information storage and retrieval system, without permission in writing from the publisher or under licence from the Copyright Licensing Agency Limited. Further details of such licences (for reprographic reproduction) may be obtained from the Copyright Licensing Agency Limited, of 90 Tottenham Court Road, London W1P 9HE.

Designed and produced by Gecko Ltd, Bicester, Oxon
Printed in Spain for Hodder & Stoughton Educational, a division of Hodder Headline Plc,
338 Euston Road, London NW1 3BH, by Graphycems.

Project manager: Jo Kemp
Mind Maps: Vanda North
Illustrations: Peter Bull, Simon Cooke, Chris Etheridge,
  Ian Law, Joe Little, Andrea Norton, Mike Parsons,
  John Plumb, Dave Poole, Chris Rothero, Anthony Warne
Cover design: Amanda Hawkes
Cover illustration: Paul Bateman

Contents

Mind Maps

1	**Telecommunications**	**12**
1.1	Communication using waves	12
1.2	Communication using cables	15
1.3	Radio and TV	18
1.4	Transmission of radio waves	22
2	**Using electricity**	**28**
2.1	From the wall socket	28
2.2	Alternating and direct current	31
2.3	Resistance	34
2.4	Useful circuits	39
2.5	Behind the wall	42
2.6	Movement from electricity	45
3	**Health physics**	**50**
3.1	The use of thermometers	50
3.2	Using sound	51
3.3	Light and sight	53
3.4	Using the spectrum	59
3.5	Nuclear radiation	61
4	**Electronics**	**70**
4.1	Overview	70
4.2	Output devices	71
4.3	Input devices	73
4.4	Digital processes	79
4.5	Analogue processes	85
5	**Transport**	**90**
5.1	On the move	90
5.2	Forces at work	93
5.3	Movement means energy	95
6	**Energy matters**	**102**
6.1	Supply and demand	102
6.2	Generation of electricity	103
6.3	Source to consumer	106
6.4	Heat in the home	111
7	**Space physics**	**121**
7.1	Signals from space	121
7.2	Space travel	126

Answers	**134**
Electrical and electronic symbols	**139**
Rearrangement of the formulae	**140**
Physical quantities – their symbols and units	**141**
Index	**142**

Revision made easy

The four pages that follow contain a gold mine of information on how you can achieve success both at school and in your exams. Read them and apply the information, and you will be able to spend less, but more efficient, time studying, with better results. If you already have another Hodder & Stoughton revision guide, skim-read these pages to remind yourself about the exciting new techniques the books use, then move ahead to page 12.

This section gives you vital **information** on how to remember more *while* you are learning and how to remember more *after* you have finished studying. It explains

- how to use special techniques to improve your memory
- how to use a revolutionary note-taking technique called Mind Maps that will double your memory and help you to write essays and answer exam questions
- how to read everything faster while at the same time improving your comprehension and concentration

All this information is packed into the next four pages, so make sure you read them!

Your *amazing* memory

There are five important things you must know about your brain and memory to revolutionise your school life.

1. how your memory ('recall') works *while* you are learning
2. how your memory works *after* you have finished learning
3. how to use Mind Maps – a special technique for helping you with all aspects of your studies
4. how to increase your reading speed
5. how to zap your revision

1 Recall during learning – the need for breaks

When you are studying, your memory can concentrate, understand and remember well for between 20 and 45 minutes at a time. Then it *needs* a break. If you carry on for longer than this without one, your memory starts to break down! If you study for hours non-stop, you will remember only a fraction of what you have been trying to learn, and you will have wasted valuable revision time.

So, ideally, *study for less than an hour*, then take a five- to ten-minute break. During the break listen to music, go for a walk, do some exercise, or just daydream. (Daydreaming is a necessary brain-power booster – geniuses do it regularly.) During the break your brain will be sorting out what it has been learning, and you will go back to your books with the new information safely stored and organised in your memory banks. We recommend breaks at regular intervals as you work through the revision guides. Make sure you take them!

2 Recall after learning – the waves of your memory

What do you think begins to happen to your memory straight *after* you have finished learning something? Does it immediately start forgetting? No! Your brain actually *increases* its power and carries on remembering. For a short time after your study session, your brain integrates the information, making a more complete picture of everything it has just learnt. Only then does the rapid decline in memory begin, and as much as 80 per cent of what you have learnt can be forgotten in a day.

However, if you catch the top of the wave of your memory, and briefly review (look back over) what you have been revising at the correct time, the memory is stamped in far more strongly, and stays at the crest of the wave for a much longer time. To maximise your brain's power to remember, take a few minutes and use a Mind Map to review what you have learnt at the end of a day. Then review it at the end of a week, again at the end of a month, and finally a week before the exams. That way you'll ride your memory wave all the way to your exam – and beyond!

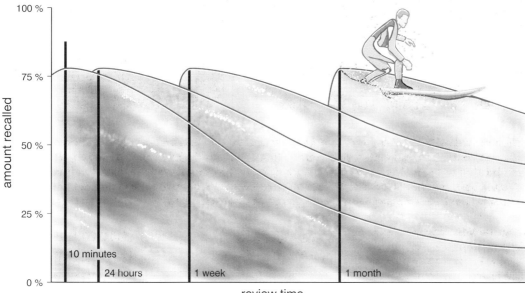

Amazing as your memory is (think of everything you actually have stored in your brain at this moment) the principles on which it operates are very simple: your brain will remember if it (a) has an image (a picture or a symbol); (b) has that image fixed and (c) can link that image to something else.

3 The Mind Map® – a picture of the way you think

Do you *like* taking notes? More importantly, do you like having to go back over and learn them before exams? Most students I know certainly do not! And how do you take your notes? Most people take notes on lined paper, using blue or black ink. The result, visually, is *boring*! And what does your brain do when it is bored? It turns off, tunes out, and goes to sleep! Add a dash of colour, rhythm, imagination, and the whole note-taking process becomes much more fun, uses more of your brain's abilities, *and* improves your recall and understanding.

A Mind Map mirrors the way your brain works. It can be used for note-taking from books or in class, for reviewing what you have just studied, for revising, and for essay planning for coursework and in exams. It uses all your memory's natural techniques to build up your rapidly growing 'memory muscle'.

You will find Mind Maps on pages 8–11 this book. Study them, add some colour, personalise them, and then have a go at drawing your own – you'll remember them far better! Put them on your walls and in your files for a quick-and-easy review of the topic.

How to draw a Mind Map

1. Start in the middle of the page with the page turned sideways. This gives your brain the maximum room for its thoughts.

2. Always start by drawing a small picture or symbol. Why? Because a picture is worth a thousand words to your brain. And try to use at least three colours, as colour helps your memory even more.

3. Let your thoughts flow, and write or draw your ideas on coloured branching lines connected to your central image. These key symbols and words are the headings for your topic. The Mind Map at the top of the next page shows you how to start.

4. Then add facts and ideas by drawing more, smaller, branches on to the appropriate main branches, just like a tree.

5. Always print your word clearly on its line. Use only one word per line. The Mind Map at the foot of the next page shows you how to do this.

6. To link ideas and thoughts on different branches, use arrows, colours, underlining, and boxes.

How to read a Mind Map

1. Begin in the centre, the focus of your topic.

2. The words/images attached to the centre are like chapter headings, read them next.

3. Always read out from the centre, in every direction (even on the left-hand side, where you will have to read from right to left, instead of the usual left to right).

Using Mind Maps

Mind Maps are a versatile tool – use them for taking notes in class or from books, for solving problems, for brainstorming with friends, and for reviewing and revising for exams – their uses are endless! You will find them invaluable for planning essays for coursework and exams. Number your main branches in the order in which you want to use them and off you go – the main headings for your essay are done and all your ideas are logically organised!

Revision made easy

4 Super speed reading

It seems incredible, but it's been proved – the faster you read, the more you understand and remember! So here are some tips to help you to practise reading faster – you'll cover the ground more quickly, remember more, *and* have more time for revision!

★ First read the whole text (whether it's a lengthy book or an exam paper) very quickly, to give your brain an overall idea of what's ahead and get it working. (It's like sending out a scout to look at the territory you have to cover – it's much easier when you know what to expect!) Then read the text again for more detailed information.

★ Have the text a reasonable distance away from your eyes. In this way your eye/brain system will be able to see more at a glance, and will naturally begin to read faster.

★ Take in groups of words at a time. Rather than reading 'slowly and carefully' read faster, more enthusiastically. Your comprehension will rocket!

★ Take in phrases rather than single words while you read.

★ Use a guide. Your eyes are designed to follow movement, so a thin pencil underneath the lines you are reading, moved smoothly along, will 'pull' your eyes to faster speeds.

5 Helpful hints for exam revision

Start to revise at the beginning of the course. Cram at the start, not the end and avoid 'exam panic'!

Use Mind Maps throughout your course, and build a Master Mind Map for each subject – a giant Mind Map that summarises everything you know about the subject.

Use memory techniques such as mnemonics (verses or systems for remembering things like dates and events, or lists).

Get together with one or two friends to revise, compare Mind Maps, and discuss topics.

And finally...

★ *Have fun while you learn* – studies show that those people who enjoy what they are doing understand and remember it more, and generally do it better.

★ *Use your teachers* as resource centres. Ask them for help with specific topics and with more general advice on how you can improve your all-round performance.

★ *Personalise your revision guide* by underlining and highlighting, by adding notes and pictures. Allow your brain to have a conversation with it!

Your brain is an amazing piece of equipment – learn to use it, and you, like thousands of students before you, will be able to master the top grades with ease. The more you understand and use your brain, the more it will repay you!

Mind Maps

Telecommunications (Unit 1)

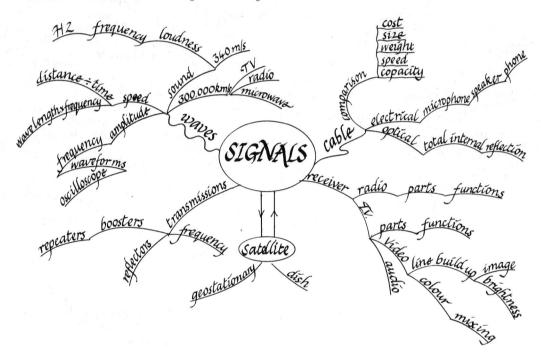

Using electricity (Unit 2)

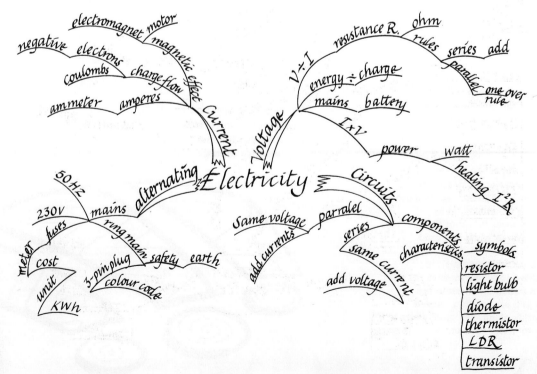

Mind Maps

Health physics (Unit 3)

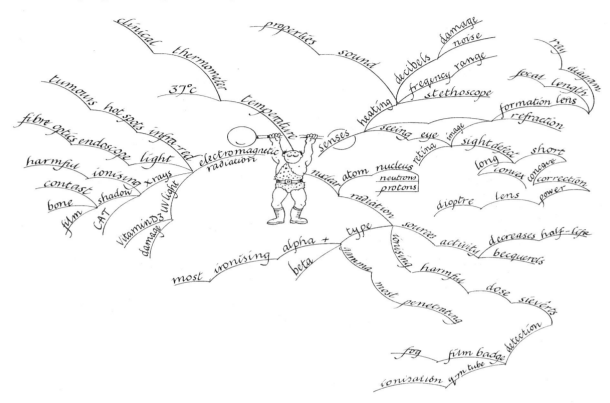

Electronics (Unit 4)

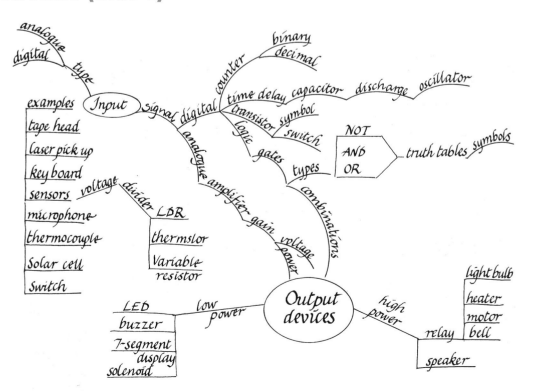

Transport (Unit 5)

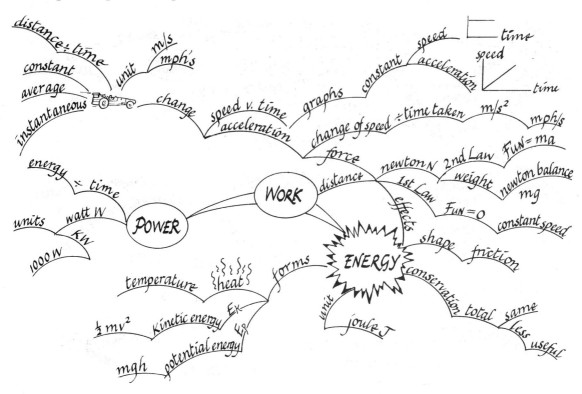

Energy matters (Unit 6)

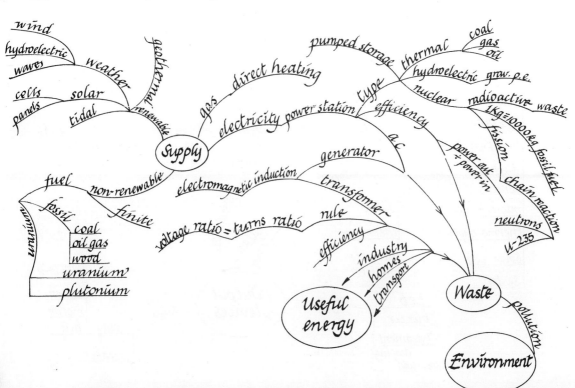

Mind Maps

Space physics (Unit 7)

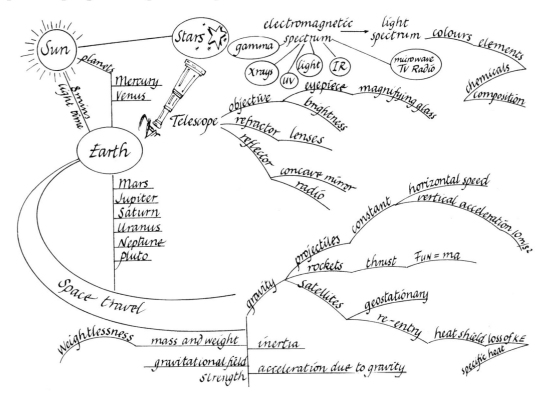

Telecommunications 1

MIND MAP Page 8.

1.1 Communication using waves

preview

At the end of this topic you will be able to:

- give an example which illustrates that the speed of sound in air is less than the speed of light in air, e.g. thunder and lightning
- describe a method of measuring the speed of sound in air (using the relationship between distance, time and speed)
- carry out calculations involving the relationship between distance, time and speed in problems on sound transmission
- state that waves are one way of transmitting signals
- use the following terms correctly in context: 'wave', 'frequency', 'wavelength', 'speed', 'energy (transfer)' and 'amplitude'
- carry out calculations involving the relationship between distance, time and speed in problems on water waves
- carry out calculations involving the relationship between speed, wavelength and frequency for water and sound waves
- explain the equivalence of $f\lambda$ and d/t.

Sound and light

The speed of sound in air is 340 m/s – about $\frac{1}{3}$ km per second.

The speed of light in air is 300 000 000 m/s (3×10^8 m/s) or 300 000 km per second.

This difference in speed explains why the action of kicking a football, observed some distance away, is heard a little time after the football is seen to leave the kicker's boot. Another good example is thunder and lightning. Both the thunder and the lightning occur at the same time but the thunder is heard some time after the lightning is seen. How far away a storm centre is can be found by timing how long the thunder takes to reach an observer's ears after the lightning is seen, e.g. a 5 second delay means the storm is 5×340 m away (=1700 m).

Measurement of the speed of sound

There are many ways of measuring the speed of sound in air using the formula:

$$\text{speed (in m/s)} = \frac{\text{distance (in m)}}{\text{time (in s)}}$$

Two people stand a measured distance apart, at least 500 m. The starter makes a loud short sound, for example with a large gong, in view of the other person (the timer). The timer times the interval between observing the sound being created and hearing it. The speed of sound is calculated from the distance divided by the time taken.

starter — AT LEAST 500 m — person with stop watch

To account for wind speed, the two people change roles and the experiment is repeated to give an average value.

One accurate method which does *not* rely on hand timing (hence *no* reaction time) is to use two microphones connected to an electronic 'fast' timer.

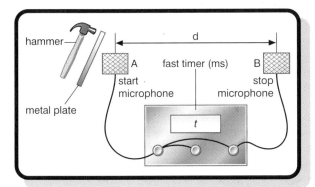

Speed of sound apparatus

A hammer hits a metal plate and the resulting sound wave starts the electronic timer as it passes microphone A. When the sound reaches microphone B, the electronic timer stops. The timer thus displays how long the sound took to travel from A to B. The distance AB between the microphones is measured with a metre stick and the speed of sound is calculated from:

$$\text{speed of sound in air} = \frac{\text{distance between microphones A and B (in m)}}{\text{time on electronic timer (in s)}}$$

Worked example

Assuming that the speed of sound in air is 340 m/s, how long does sound take to travel 1 m?

Solution
From the triangle,
$t = \frac{d}{v} = \frac{1}{340} = 0.0029$ s

Worked example

During a fireworks display a flash of light is seen in the night sky and 2 seconds later a bang is heard. How far away is the firework explosion?

Solution
From the triangle,
$d = v \times t = 340 \times 2 = 680$ m

Transmitting messages using waves

Radio waves, TV waves and microwaves can be used to carry messages or signals from a transmitter to a receiver. These waves travel at the speed of light (300 000 000 m/s) and are particularly useful where the distances over which the messages have to be carried are large.

What is a wave?

Waves can carry energy without carrying matter.
Drop a stone in a pond and observe the ripples as they spread out. A small object floating on the water would bob up and down as the ripples pass it. The ripples are waves on the water surface carrying energy across the pond. A water wave is an example of a disturbance in a substance which travels through the substance. Electromagnetic waves do not need a substance to travel through; all other types of waves do.

Facts about waves

★ **The amplitude of a wave** is the height of the wave crest above the centre. This is the same as the depth of a wave trough below the centre.

★ **The wavelength of a wave** (symbol λ, pronounced 'lambda') is the distance from one crest to the next crest. This is the same as the distance from a trough to the next trough.

★ **The frequency of a wave** (symbol f) is the number of crests passing a given position each second. This is the same as the number of complete waves per second passing a fixed point. The unit of frequency is the hertz (symbol Hz), equal to 1 cycle per second.

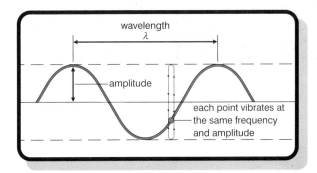

Revise Standard Grade Physics

Speed

★ The speed v of a wave is the distance travelled by a crest in one second. The unit of speed is the metre per second (m/s).

★ The speed of a wave can be calculated from its frequency and wavelength using the following equation:

speed = frequency × wavelength
(in m/s)   (in Hz)   (in m)

Worked example

Water waves pass a fixed point at the rate of 20 waves every 10 seconds. If the distance between successive crests is 1.5 m, what is the wave speed?

$f = \frac{20}{10} = 2$ waves per second = 2 Hz

$\lambda = 1.5$ m

$v = f\lambda$
$= 2 \times 1.5$
$= 3$ m/s

Worked example

A tuning fork gives out sound waves of frequency 440 Hz. What is the wavelength of the sound waves?

Solution

From the triangle,

$\lambda = \frac{v}{f} = \frac{340}{440} = 0.77$ m

The equivalence of $v = f\lambda$ and $v = d/t$

When a wave moves a distance of one wavelength (i.e. d equals λ), the time taken is $1/f$ seconds. To show that this is the case, imagine a wave of frequency 10 Hz. This means there are 10 waves every second, so there will be one wave every $\frac{1}{10}$ th of a second. In other words the time taken for the wave to move one wavelength is $\frac{1}{10}$ th of a second (i.e. $1/f$ seconds),

therefore $v = \frac{d}{t}$ becomes $v = \frac{\lambda}{1/f}$ i.e. $v = f\lambda$

hence $v = \frac{d}{t}$ is equivalent to $v = f\lambda$

and either can be used to find the wave speed depending on the information given.

Questions

1 What are the wavelength and the amplitude of the wave shown on page 13?

2 Complete the table below by calculating the missing values of speed, frequency or wavelength.

wavelength/m	2.5	0.1			5×10^{-7}
frequency/Hz		3400	500	3750	
speed/m/s	340		80	1500	3×10^8

3 Suppose the wave in the diagram on page 13 is travelling at a speed of 60 mm/s to the left. Calculate its frequency using the wavelength measurement from question 1.

Answers
1 Amplitude = 10 mm, wavelength = 40 mm
2 136 Hz, 340 m/s, 0.160 m, 0.40 m, 6×10^{14} Hz
3 2.4 Hz

1.2 Communication using cables

preview

At the end of this topic you will be able to:
- describe a method of sending a message using code (Morse or similar)
- state that coded messages or signals are sent out by a transmitter and are picked up by a receiver
- state that the telephone is an example of long-range communication between transmitter and receiver
- state the energy changes in (a) a microphone (sound to electrical) and (b) a loudspeaker (electrical to sound)
- state that the mouthpiece of a telephone (transmitter) contains a microphone and the earpiece (receiver) contains an earphone
- state that electrical signals are transmitted along wires during a telephone communication
- state that an electrical signal is transmitted along wires at a speed much greater than the speed of sound (almost 300 000 000 m/s)
- describe the effect on the signal pattern displayed on an oscilloscope due to a change in (a) the loudness of sound and (b) the frequency of sound
- describe, with examples, how the following terms relate to sound: 'frequency' and 'amplitude'
- state what is meant by an optical fibre
- describe one practical example of telecommunication which uses optical fibres
- state that electrical cables and optical fibres are used in some telecommunication systems
- state that light can be reflected
- describe the direction of the reflected light ray from a plane mirror
- state that signal transmission along an optical fibre takes place at very high speed
- explain the electrical signal pattern in telephone wires in terms of loudness and frequency changes in the sound signals
- compare some of the properties of electrical cables and optical fibres, e.g. cost, size, weight, signal speed, signal capacity, signal quality and signal reduction per kilometre
- state the principle of reversibility of ray paths
- describe the principle of operation of an optical fibre transmission system
- carry out calculations involving the relationship between distance, time and speed in problems on light transmission

Morse code telegraph

This device sends coded messages along wires as pulses of electricity. The pulses are produced by opening and closing a switch in a simple circuit consisting of a battery, a switch and a buzzer or bulb. There are two kinds of pulses, short pulses and long pulses, and there are different combinations of long and short pulses for each letter of the alphabet.

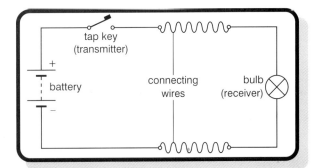

Morse code telegraph

In the telegraph circuit, the switch is the **transmitter** and the buzzer or bulb is the **receiver**.

The telephone

One of the most common methods of long-range communication using wires is the telephone. The mouthpiece of a telephone (transmitter) contains a microphone, which converts sound energy to electrical energy. The electrical signals travel down the wire at almost the speed of light. The earpiece (receiver) contains a loudspeaker, which transforms the electrical energy back into sound energy.

Revise Standard Grade Physics

Sound patterns

Pure sounds picked up by a microphone can be displayed on an oscilloscope. The oscilloscope trace shows whether a sound is loud or soft, or has a low or a high pitch.

The **louder** the sound, the greater the **amplitude** of the wave pattern.

The **higher** the pitch, the greater the number of waves, i.e. the greater the **frequency**.

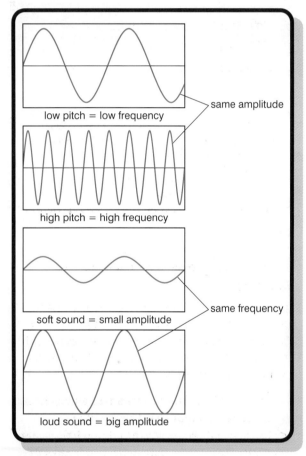

Sound pictures

Optical fibre

This consists of a very thin strand of pure glass. Light pulses can be transmitted down an optical fibre at very high speed. Optical fibres are usually grouped together into a thicker optical cable and are used extensively in the telecommunications network to transmit telephone and TV signals.

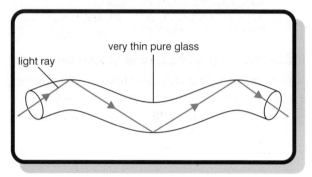

Optical fibre

Light reflection

When light hits a plane mirror, it is reflected back such that the angle of incidence equals the angle of reflection.

The **law of reflection** states that the angle between the reflected ray and the normal is equal to the angle between the incident ray and the normal.

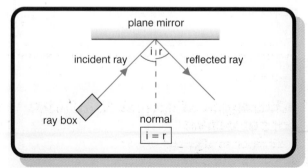

The law of reflection

Questions

1 What is the function of **a)** a microphone, **b)** a loudspeaker?

2 An oscilloscope is used to display sound waveforms as shown above. Sketch the waveform you would expect for:
   a) a soft sound at high pitch, in comparison with waveform 1
   b) the same loudness as waveform 4 but at a lower pitch.

Answers

1 a) A microphone converts sound waves to electrical waves. **b)** A loudspeaker converts electrical waves to sound waves.
2 a) The waves would be smaller in height and closer together. **b)** The waves would be at the same height but stretched out more.

The diagram shows how this can be tested using a ray box and a plane mirror. Note that the normal is the line which is perpendicular to the mirror at the point where the light ray meets the mirror.

Question

3 a) In the diagram of the optical fibre on page 16, how many times is the light ray reflected?
b) Why does this light ray take a little longer to travel through the fibre than if the fibre was straight?

Sound and electrical patterns

When we speak into the mouthpiece (microphone) of a telephone, the sound signal is very complicated because of all the different sounds which make up each word we say. Added to this we may change pitch and loudness as the conversation develops. This will produce an identical series of changes in the electrical signal pattern in the wires.

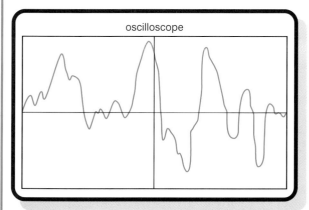

Speech trace

Advantages of optical fibres over copper cables

1. Cheaper, smaller and lighter.
2. Can carry more information.
3. Signal quality is better.
4. Signals need to be boosted less because very little energy is lost.

Reversibility of light

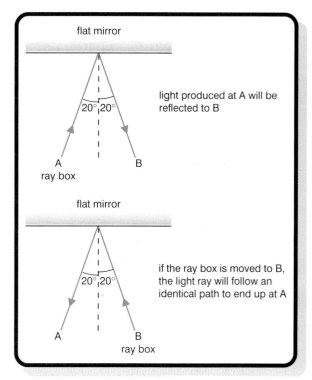

Reversibility of light

An optical fibre transmission system

When a ray of light travelling in an optical fibre hits the surface of the glass (at an angle greater than the critical angle) the light is reflected back into the fibre and does not escape. This is called **total internal reflection**.

A sound signal can be converted into an electrical signal using a microphone. The resulting electrical signal can be converted into light energy and transmitted via an optical fibre to a receiver. This receiver can change the light energy back into electrical energy to operate a loudspeaker. This is an example of a modern optical fibre transmission telephone system.

Answer 3 a) Three times. b) The points of reflection would be further apart if it was straight so it would cross from side to side less, and therefore travel through the fibre more quickly.

Optical fibre calculations

Worked example

Note that $v = d/t$ can be applied to optical fibres, where $v = 200\,000\,000$ m/s (or 2×10^8 m/s).

How long will it take light to travel down an optical fibre 25 km long?

Solution

From the speed, distance, time triangle,

$t = \frac{d}{v} = \frac{25\,000}{2 \times 10^8} = 1.25 \times 10^{-4}$ s

Worked example

How far does light travel through an optical fibre in 4×10^{-6} s?

Solution

$d = v \times t = 2 \times 10^8 \times 4 \times 10^{-6} = 800$ m

Total internal reflection

This occurs when light in a transparent medium strikes the boundary at an angle of incidence greater than a **critical angle**. The light ray reflects internally just as if the boundary is a mirror.

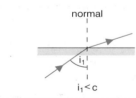

1. If the angle of incidence is less than the critical angle, the light ray bends away from the normal on leaving the glass.

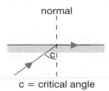

c = critical angle

2. If the angle of incidence is equal to the critical angle, the light refracts along the boundary.

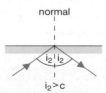

3. If the angle of incidence exceeds the critical angle, the light ray is totally internally reflected.

An optical fibre is a thin fibre of transparent flexible material. A light ray that enters the fibre at one end emerges at the other end, even if the fibre is curved round. This happens because the light ray is totally internally reflected at the fibre surface wherever it hits the boundary. Each light ray in the fibre travels along a straight line through the fibre between successive reflections. Provided the bends in the fibre are not too tight, light rays in the fibre do not emerge from its sides.

Questions

4 a) State two advantages of optical fibres over copper cables.
b) Why is the amount of light emerging from an optical fibre reduced if the fibre is bent too much?

5 a) The speed of light in a certain fibre is 200 000 km/s. Calculate the least time taken by light to travel along a 2 km length of this fibre.
b) Short pulses of light are transmitted along the fibre at a rate of 1 million pulses per second. Calculate the least distance between two pulses in the fibre.

Answers

4 a) See page 17 **b)** The angle of incidence is reduced by bending the fibre. Light will escape from the sides of the fibre wherever the angle of incidence is less than the critical angle, i.e. if the fibre is bent too much.

5 a) 10 μs **b)** 0.2 km

1.3 Radio and TV

preview

At the end of this topic you will be able to:

- state that the main parts of a radio receiver are the aerial, tuner, decoder, amplifier, loudspeaker, electricity supply and identify these parts on a block diagram
- describe in a radio receiver the function of the aerial, decoder, amplifier, loudspeaker and electricity supply
- state that the main parts of a TV receiver are the aerial, tuner, decoders, amplifiers, tube, loudspeaker, electricity supply and identify these parts on a block diagram of a TV receiver

Telecommunications

- describe in a TV receiver the function of the aerial, tuner, decoders, amplifiers, tube, loudspeaker and electricity supply
- describe how a picture is produced on a TV screen in terms of line build-up
- state that mixing red, green and blue lights produce all the colours seen on a colour TV screen
- describe the general principles of radio transmission in terms of transmitter, carrier wave, amplitude modulation and receiver
- describe the general principle of TV transmission in terms of transmitter, carrier wave, modulation, video receivers and audio receivers
- describe how a moving picture is seen on a TV screen in terms of line build-up, image retention and brightness variation
- describe the effect of colour-mixing lights (red, green and blue).

Radio receiver

The six main parts of a radio receiver are:

aerial → tuner → decoder → amplifier → loudspeaker
                                ↑
                          electricity supply

This is called a **block diagram**.

The function of each of the six main parts is described below.

Aerial: this is made of metal and picks up the radio waves coming from transmitters and turns them into electrical signals.

Tuner: this picks out *one* radio frequency or station from the many different frequencies arriving at the aerial.

Decoder: this separates the music or speech (i.e. the low frequency audio wave) from the high frequency radio wave.

Amplifier: this increases the size of the electrical signal before it is sent to the loudspeaker.

Loudspeaker: this changes electrical energy into sound energy.

Electrical supply: this gives the amplifier the energy it requires to make it operate. This can be mains electricity or batteries.

TV receiver

A block diagram for a TV receiver is shown. Notice the similarity with a radio receiver.

The aerial, tuner, decoders, amplifiers, electrical supply and loudspeaker have the same function as in a radio receiver. The TV tube changes electrical energy into light energy to make the picture.

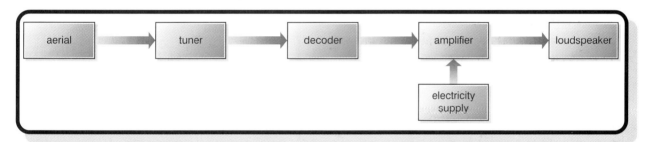

Producing a black and white picture

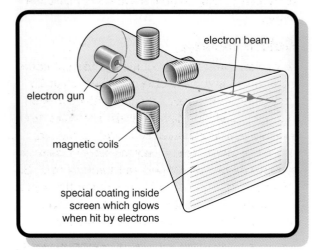

Section through a TV tube

The picture tube has nearly all the air sucked out of it. An electron gun fires a beam of electrons towards the screen. Magnetic coils move the electron beam up, down or sideways, so the electrons can be aimed at any part of the screen. The screen is coated on the inside with a special material which glows when it is hit by electrons. The more electrons which hit a certain area of the screen, the brighter it is. In other words, light energy is produced only when electrons hit the screen.

The picture is built up by moving the electron beam backwards and forwards across the screen whilst moving it down the screen to create 625 lines per picture.

Colour TV

All the colours we see on the screen of a TV tube can be made by combining different brightnesses of the three colours of light, namely **red**, **green**, and **blue**. To achieve this, there are three electron guns in a colour TV tube.

Questions

1 In a radio receiver, what is the function of **a)** the aerial **b)** the amplifier?

2 a) How many electron guns are present in **(i)** a black and white TV tube **(ii)** a colour TV tube?

b) (i) How many lines are scanned by each electron beam in a colour TV tube as it moves from the top to the bottom of the screen?

(ii) How would the picture on a colour TV tube be affected if two of the electron guns failed?

Radio transmission

Low frequency **audio waves**, which we can hear, cannot travel very far. Very high frequency **radio waves**, however, can travel very long distances. If these two types of waves are mixed together, an **amplitude modulated wave** is produced, which can be transmitted through the air to a radio receiver.

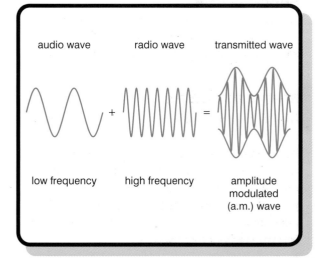

Amplitude modulated wave

The radio wave in this process is called the **carrier wave**. The transmitted wave can be decoded at the radio receiver to separate the sound (audio) wave from the radio wave. More commonly nowadays, audio and radio waves are combined together to give **frequency modulated** (FM) waves, which give better reception and suffer less from interference.

Answers

1 a) The aerial picks up radio waves from the transmitter and turns them into electrical signals. **b)** The amplifier increases the size (i.e. amplitude) of the electrical signal.
2 a) (i) 1 **(ii)** 3 **b) (i)** 625 **(ii)** The picture would be one colour only.

TV transmission

The same principles are used for TV transmission and radio transmission, except that for TV transmission both video (picture) signals and audio (sound) signals are required (see diagram below).

A TV picture

There are three factors to consider when explaining the production of a moving picture:

1. line build-up
2. image retention
3. brightness variation.

Line build-up

625 lines per picture are created by an electron beam scanning across the TV tube. This happens so quickly that the eye cannot detect the individual images.

Image retention

25 different pictures are produced each second and so there is a gap of 1/25th of a second between pictures. The human eye keeps (retains) an image for a short period of time (approximately 1/10th of a second) after the picture has disappeared. This image retention, or persistence of vision, ensures that the separate images merge into each other.

Brightness variation

The more electrons that hit the screen, the brighter the picture. Light and dark parts of a picture are made by changing the number of electrons (or intensity) in the electron beam, which in turn changes the brightness of the spot. The electron beam can be switched on or off extremely quickly to produce areas of black and white.

Colour mixing

Any colour can be made by mixing different intensities of **red, green** and **blue** light. In a colour TV the inside of the screen is coated with fluorescent red, green and blue dots or pixels. There are three electron guns each aimed separately at the red, green and blue pixels. If only the red gun is on, then it will hit only red dots and similarly with the green and blue guns. If, however, two or three guns are on, then mixing of the colours takes place, which produces more colours as shown below.

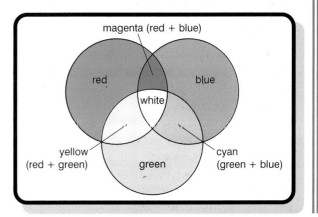

Colour mixing

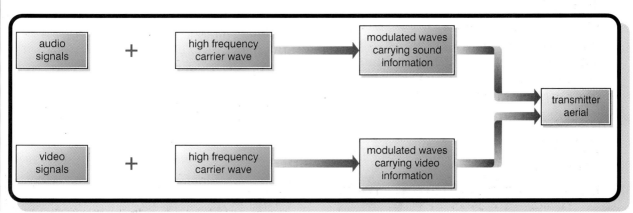

TV transmission

Revise Standard Grade Physics

Questions

3 a) What is meant by the amplitude modulation of a carrier wave?
b) State one advantage of frequency modulation in comparison with amplitude modulation.
4 a) (i) How many pictures are produced each second on the screen of a colour TV?
  (ii) What difference would the viewer see if the number of pictures per second was reduced to 10?
b) How would the picture on a colour TV tube be affected if one of the electron guns failed?

Answers

3 a) Amplitude modulation of a carrier wave is the variation in its amplitude according to the amplitude of the signal being carried.
b) Better reception, less interference.
4 a) (i) 25 **(ii)** The pictures on the screen would flicker.
b) One of the three primary colours would disappear so the picture would be in the complementary colour of that primary colour.

1.4 Transmission of radio waves

preview

At the end of this topic you will be able to:

- state that mobile telephones, radios and TVs are examples of long-range communication applications that do not need cables (between transmitter and receiver)
- state that microwaves, TV signals and radio signals are waves that transfer energy
- state that microwaves, TV signals and radio signals are transmitted at very high speeds
- state that microwaves, TV signals and radio signals are transmitted through air at 3×10^8 m/s
- state that a radio transmitter can be identified by wavelength or frequency values
- state that curved reflectors on certain aerials or receivers make the received signal stronger
- explain why curved reflectors on certain aerials or receivers make the received signal stronger
- describe an application of curved reflectors used in, telecommunication, e.g. satellite TV, TV link boosters, repeaters or satellite communication
- state that the period of a satellite orbit depends on its height above the Earth
- state that a geostationary satellite stays above the same point on the Earth's surface
- describe the principle of transmission and reception of satellite TV broadcasting using geostationary satellites and dish aerials
- describe the principle of intercontinental telecommunication using a geostationary satellite and ground stations
- carry out calculations involving the relationship between distance, speed and time in problems concerning microwaves, TV waves and radio waves
- carry out calculations involving the relationship between speed, wavelength and frequency for microwaves, TV waves and radio waves
- explain some of the differences in properties of radio bands in terms of source strength, reflections, etc.
- explain in terms of diffraction how wavelength affects radio and TV reception
- explain the action of curved reflectors on certain transmitters.

Long-range communication

Energy can be transferred from a transmitter to a receiver in the form of waves that do not require cables/wires to carry them. Radio waves, TV waves and microwaves are examples of such waves and they travel through the air at 3×10^8 m/s or 300 000 000 m/s (the same as the speed of light). Radios, TVs and mobile phones are examples of long-range communication applications that do not require wires.

Frequency bands

	frequency range	uses
long wave (LW)	up to 300 kHz	international AM radio
medium wave (MW)	300 kHz–3 MHz	AM radio
high frequency (HF)	3–30 MHz	AM radio
very high frequency (VHF)	30–300 MHz	FM radio
ultra high frequency (UHF)	300–3000 MHz	TV broadcasting, mobile phones
microwave	above 3000 MHz	satellite TV, global phone links
light	500 THz approx	fibre optic communication links

Note: 1 MHz = 1 000 000 Hz. 1 THz = 1 million MHz.

Identification of a radio transmitter

A radio transmitter (station) can be identified in terms of the frequency and wavelength of the signal it transmits or broadcasts, e.g. Atlantic 252 has a wavelength of 252 m and Radio Forth has a frequency of 97.3 MHz. These values are unique to the particular station. Since all radio waves travel at 3×10^8 m/s, then specifying a frequency automatically specifies a particular wavelength.

Mobile phone links are only possible if the receiver is near a transmitter. A mobile phone signal is allocated a channel of bandwidth 25 kHz in the UHF band at a frequency of about 900 MHz. The transmitter is linked to the international phone network which uses undersea cable links, local microwave links and satellite links.

Radio broadcasts at frequencies below 30 MHz travel long distances due to reflection from a layer of ionised gases in the upper atmosphere. Long wave broadcasts spread round the Earth because long wavelength radio waves follow the Earth's curvature.

Satellite TV signals are carried by microwaves from a geostationary satellite. This orbits the Earth once every 24 hours round the equator, so stays in the same place above the Earth. A reflecting dish pointed towards the satellite focuses the microwaves on an aerial, which detects the signal and passes it onto a decoder.

Terrestrial TV signals are carried by radio waves in the UHF band range. Receiving aerials need to be in the line of sight of the transmitter. TV pictures from the other side of the world reach us via satellite links and ground stations.

Curved reflectors

Satellite dishes and aerials are curved to collect weak signals and focus them onto a receiver placed at the focus. Since the signal is concentrated at the focus of the curved reflector, it is much stronger.

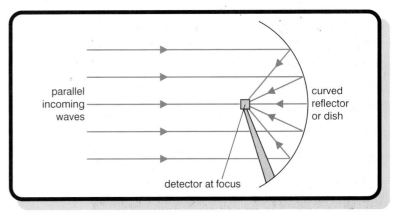

Curved reflector as a receiver

Curved reflectors are used to collect signals from satellites, which are used as TV links and as microwave repeaters, to boost signals at regular intervals.

If a transmitter is placed at the focus, then a parallel beam of waves is sent out by the transmitting aerial, which is the above in reverse.

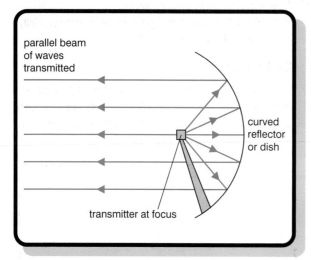

Curved reflector as a transmitter

Satellite communications

Telecommunication satellites move round the Earth in an **orbit** gathering information or relaying it from a transmitter to a receiver. The time taken to complete one orbit is called the **period** and depends on the height of the satellite above the Earth. The higher the satellite, the great the period.

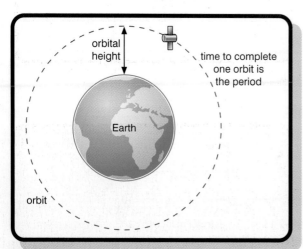

Satellite in orbit

A geostationary satellite orbits at a height of approximately 36 000 km above the equator and has a period of 24 hours. Since the time taken for the Earth to revolve is also 24 hours, it appears to be stationary above the same point on the Earth's surface. Satellite TV broadcasts use geostationary satellites to receive, amplify and transmit signals to different parts of the Earth's surface. Curved dish aerials on the satellites are used to both receive and transmit signals.

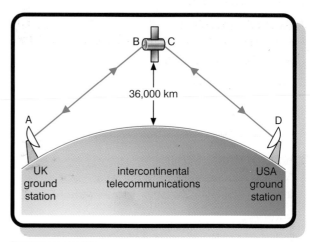

Geostationary satellite

A: dish aerial transmits signal to curved reflector B on geostationary satellite

B: signal received and amplified

C: signal re-transmitted by a dish aerial to receiving aerial at ground station D

D: received signal brought to a focus on a curved receiving dish aerial and a strong signal picked up by detector at the focus

By the same method, a signal can be sent from D to A.

> **Questions**
>
> **1 a)** State the speed of radio waves through air.
> **b)** A certain radio station X broadcasts on 101.2 MHz. Another radio station Y broadcasts on 102.0 MHz. What is the difference between the two station frequencies in kHz?
> **c)** Why is it necessary to allocate different carrier frequencies to radio stations in the same region?

2 A satellite transmits microwaves to a ground station from a dish and aerial fixed to the outside of the satellite.
  a) What is the purpose of the dish?
  b) Why must the aerial be carefully located in relation to the dish?

Calculations with waves using $v = d/t$ or $v = f\lambda$

Microwaves, radio waves and TV waves all travel at a speed of 3×10^8 m/s, which is the same as the speed of light in air.

Worked example 1
How long will it take radio waves to travel 40 km from a transmitter?

Solution
$v = \frac{d}{t}$

$t = \frac{d}{v} = \frac{40000}{3 \times 10^8} = 1.33 \times 10^{-4}$ s

Worked example 2
A radio station transmits radio waves at a frequency of 300 kHz. What is their wavelength?

Solution
$v = f\lambda$

$\lambda = \frac{v}{f} = \frac{3 \times 10^8}{3 \times 10^5} = 1000$ m

Properties of radio bands

Band	Properties
LW and MW	Strong signals. Unaffected by tall obstacles and buildings. Follow the curvature of the Earth
SW	Reflected off the ionosphere Can travel world-wide
VHF and UHF	Line of sight reception only Reflected off hills

Questions

3 a) Calculate the wavelength of radio waves of frequency 100 MHz.
  b) (i) Why can long wave broadcasts from America be picked up in Europe?
     (ii) Why is it not possible to listen to your local radio station when you are in another country?

4 a) What is a geostationary satellite?
  b) Why do satellite receiving dishes need to be carefully aligned when being set up?

Diffraction

Waves spread out when they pass through a gap or behind an obstacle. The longer the wavelength, the greater the spreading. The waves spread out more through a narrower gap.

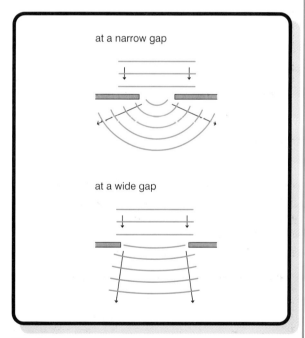

Diffraction

Answers
1 a) 300 000 km/s b) 800 kHz c) if they were operating at the same frequency, listeners would hear both stations at the same time and interference would also occur.
2 a) To reflect microwaves from the aerial into a narrow beam.
  b) The beam would spread too much if the aerial was too close to the dish and it would not spread enough if the aerial was too far from the dish.

Answers
3 a) 3.0 m b) (i) Long wave carrier waves follow the Earth's curvature. (ii) Local radio carrier waves are VHF radio waves, which follow the line of sight from the transmitter.
4 a) A satellite in a 24-hour orbit that remains directly above a point on the equator.
  b) To ensure that the waves reflected by the dish are focused on the aerial.

Revise Standard Grade Physics

Radio and TV waves travel long distances from transmitter to receiver. Occasionally there will be hills in the way which may block the signal. However, waves can bend around hills by a process known as **diffraction**.

The bigger the wavelength, the greater the amount of bending or diffraction. So in a hilly region, only long wavelengths will be received.

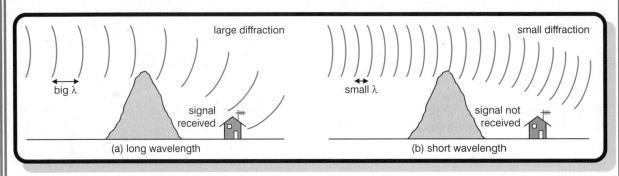

Waves bending round hills

Equations review

1. $v = \dfrac{d}{t}$ In this section v is the speed of something and the something travels at the same speed all the time

2. $v = f\lambda$ f here is the frequency of the wave, λ is 'lambda', the Greek letter used for the length of the wave (crest to crest)

Often problems require **both** equations to be used

round-up

1 The speed of sound in air is 340 m/s.
 a) Calculate the frequency of sound waves of wavelength 0.10 m in air. [1]
 b) Calculate the wavelength of sound waves of frequency 5000 Hz in air. [1]

2 The diagram shows a ship at sea near a cliff. An echo from the ship's siren is heard 5.0 s after the siren is sounded.
 a) Calculate the distance from the ship to the cliff. The speed of sound in air is 340 m/s. [2]
 b) If the ship was fog bound, explain how the ship's captain could find out if the ship was moving towards or away from the cliffs. [2]

3 A wave machine in a swimming pool vibrates 360 times every minute.
 a) What is the frequency of the waves? [2]
 b) How many waves are generated in 30 minutes? [2]

4 The diagram shows a snapshot of a wave travelling from left to right.
Use a millimetre rule to measure its wavelength and its amplitude. [2]

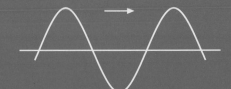

5 If the wave in question **4** is travelling at 20 mm/s, what is
  a) its frequency? [1]
  b) the number of complete waves produced in 1 minute? [1]

6 Calculate the wavelength of ultrasonic waves of frequency 1.5 MHz in water, given that the speed of the waves is 1500 m/s. [2]

7 Diagram **A** shows an oscilloscope trace produced by a sound of a certain frequency and loudness. Which of the alternative diagrams, **B, C, D** or **E** shows how the trace differs from that in diagram **A** if the sound is made
  a) more quietly at the same frequency? [1]
  b) more loudly at a lower frequency? [1]

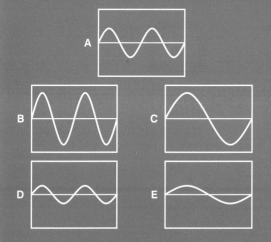

8 A light ray was totally internally reflected when it was incident on a boundary between air and glass.
  a) Was the light ray incident on air or glass? [1]
  b) Was the angle of incidence of this light ray less than, equal to or greater than the critical angle of the glass? [1]

9 How long would it take a signal to travel 100 km down an optical fibre if the speed of the signal is 2×10^8 m/s? [2]

10 Five of the parts of a radio receiver are listed alphabetically below:
  1 Aerial **2** Amplifier **3** Decoder
  4 Loudspeaker **5** Tuner
  a) List these parts in the order in which a radio signal passes through them. [5]
  b) What is the purpose of the amplifier? [1]

11 a) Why does a television receiver have two decoder circuits? [2]
  b) Why does a colour television tube have three electron guns? [1]
  c) How would a television picture differ if two of the electron guns failed? [1]

12 In a TV receiver there are 625 horizontal lines making up each picture and 25 pictures are produced each second.
  a) How many pictures are produced in 2 minutes? [2]
  b) How many horizontal lines are drawn in 2 minutes? [2]
  c) What is the time between each picture being produced? [1]

13 Explain why a house behind a hill may not receive a TV picture but may receive sound. [2]

14 a) What is meant by a geostationary satellite? [2]
  b) What is the purpose of the metal dish of a satellite receiver? [2]

15 a) Explain how it is possible to detect radio broadcasts from distant countries. [2]
  b) (i) What is meant by the carrier frequency of a radio or TV broadcast? [1]
    (ii) Why is it necessary for TV transmitter stations in adjacent regions to broadcast at different carrier frequencies? [1]
  c) Why is it necessary to use a concave metal dish to detect signals from a satellite but not from a TV mast? [2]

Total = 49 marks

Using electricity

2.1 From the wall socket

MIND MAP Page 8.

preview

At the end of this topic you will be able to:
- describe the mains supply/battery as a supply of electrical energy and describe the main energy transformations occurring in household appliances
- state the approximate power ratings of different household appliances
- select an appropriate flex given the power rating of the appliance
- state that fuses in plugs are intended to protect flexes
- select an appropriate fuse given the power rating of the appliance
- identify the live, neutral and earth wires from the colour of their insulation
- state to which pin each wire must be connected for a plug, lampholder and extension socket
- state that the human body is a conductor of electricity and that moisture increases its ability to conduct
- state that the earth wire is a safety device
- state that electrical appliances which have the double insulation symbol do not require an earth wire
- draw the double insulation symbol
- explain why situations involving electricity could result in accidents (proximity of water, wrong fuses, wrong, frayed or badly connected flexes, short circuits and misuse of multiway adapters)
- explain how the earth wire acts as a safety device
- explain why fuses and switches must be in the live lead.

Electrical energy

Many household appliances are powered by electricity. They all change electrical energy into some other form of energy. The electrical energy they require is supplied by batteries or the mains.

Some examples are:

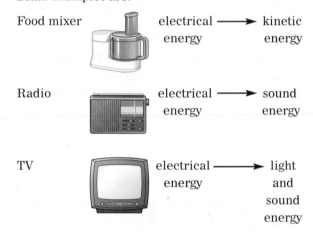

Food mixer — electrical energy → kinetic energy

Radio — electrical energy → sound energy

TV — electrical energy → light and sound energy

Power ratings

Household electrical appliances have their power ratings listed on a rating plate, which is usually underneath the appliance.

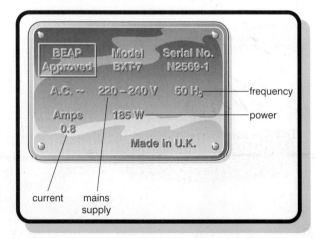

A rating plate

Power is measured in watts (W) and is equal to the rate at which electrical energy is transferred into other forms of energy. As a general rule, appliances which produce a lot of heat have high power ratings.

Using electricity

Name of appliance	Power rating (W)
Table lamp	60
Colour TV	330
Electric fire	3000

Choosing the correct flex

A flex (flexible cord) connects an electrical appliance to the mains supply by means of a plug. A flex carrying too much current could overheat, which may cause a fire. The higher the power rating, in watts (W), the thicker the flex must be. Flexes have either two or three cores (wires) depending on whether or not the appliance needs an earth wire. The cover material for the flex can be rubber or PVC, which is a type of plastic, or fabric if it needs to bend a lot or be heat resistant – an iron has a fabric-covered flex for these reasons.

Fuses within plugs

Inside a plug there is a fuse, which is intended to protect the flex of the appliance.

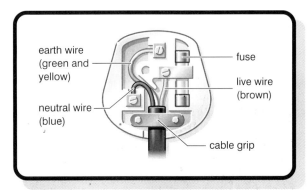

A mains plug

The fuse wire inside the cartridge is made of a tin/lead alloy, which melts or 'blows' if too high a current flows through the flex. This cuts off the supply of electrical energy.

Choosing the correct fuse

Firstly, the rating plate should be viewed to check the power rating of the appliance. Secondly, apply the following rule:

up to 700 W, use a 3 A fuse

over 700 W, use a 13 A fuse

Inside the plug

The three wires inside a plug are called **live**, **neutral** and **earth**. The colours of the live, neutral and earth wires found inside a plug are:

L = live   brown

N = neutral   blue

E = earth   green and yellow stripes

When wiring a plug or extension socket, the live, neutral and earth wires in the flex must be connected to the live, neutral and earth pins, respectively, in the plug. However, when wiring a plastic lampholder, only the live and neutral wires in the two-core flex are connected to the live and neutral pins in the lampholder.

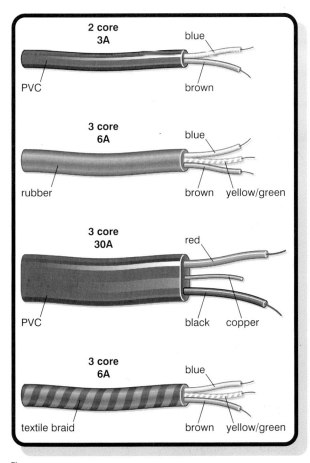

Flexes

Electrical safety

Human conductivity
Humans can conduct electricity! This is especially true if the skin is wet or damp because this makes the body an even better conductor. For this reason, humans should not touch faulty appliances as there is a risk of electrocution, and this risk increases if the skin is wet. This is why mains appliances are not allowed in bathrooms, where there is a lot of steam and water. The light switch should be either outside the bathroom or, if it is inside, there should be a pull-cord to operate the switch.

The earth wire
An electric current will flow along the earth wire in preference to other paths (i.e. through you) when a faulty appliance becomes 'live'. The earth wire is therefore a safety device which prevents the user from getting an electric shock.

Double insulation
When an appliance has the double insulation sign on its rating plate it does not need an earth wire, as all the metal parts are insulated with plastic. Such appliances, e.g. hair driers are called **double insulated** and have only live and neutral wires in their flexes.

Double insulation symbol

Faults and fuses

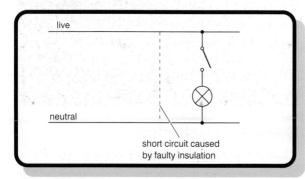
A short circuit

A **short circuit** occurs where a fault creates a low resistance path between two points at different voltages. The current through the short circuit is much greater than the current along the correct path between the two points, enough to create a fire through overheating.

Fuses are intended to prevent excessive currents flowing. A fuse is a thin piece of resistance wire which overheats and melts if too much current passes through it. The fuse wire breaks when it melts, thus cutting the current off and protecting the appliance or the wires leading to it from overheating due to excessive current.

Faults in mains circuits can arise due to

- **poor maintenance** e.g. frayed cables or damaged plugs or fittings such as sockets and switches
- **carelessness** e.g. cables that are too long, coiling a cable (which prevents heat from escaping from it)
- **overloading a circuit** e.g. too many appliances connected to the same circuit or connecting a powerful appliance to a low current circuit.

Questions

1. In a three-pin plug, what is the colour of **a)** the neutral wire **b)** the live wire **c)** the earth wire?
2. Given a 3 A fuse and a 13 A fuse, which one would you fit to **a)** a 3000 W electric kettle **b)** a 330 W colour TV?

Dangerous situations
Electricity can be dangerous unless a few basic rules are followed:

DO
1. use the correct size of fuse
2. consult an electrician if in doubt
3. wire plugs correctly
4. check that flexes are not worn, frayed or cut
5. check that plugs are not cracked
6. check that sockets are not cracked

DO NOT
1. handle appliances with wet hands
2. connect too many appliances into an adapter - this can draw too much current from the mains and cause overheating

Accidents such as electrocution or fires may occur unless great care is taken when electricity is being used.

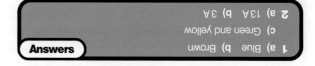
Answers: 1 a) Blue b) Brown c) Green and yellow 2 a) 13A b) 3A

Earth wire as a safety device

If an electrical appliance that is not earthed develops a fault and the live wire touches the metal casing, then anyone touching the casing will get an electric shock. The electric wire is connected to the metal casing so the current will pass through the earth wire to earth instead of passing through the person touching it. The current in the earth wire may blow the fuse and so cut off the electrical supply. However, the current through the fuse may not be sufficient to blow it and therefore the appliance would be lethal if the earth connection was broken.

Switches and fuses

The switch and the fuse must be placed in the live wire as shown.

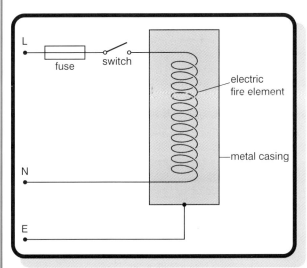

Circuit with switch and fuse

If the switch and fuse were in the neutral wire, the fire would still work but even if the fire were switched off or the fuse had blown, the element would still be connected directly to the live wire, making it extremely unsafe.

2.2 Alternating and direct current

preview

At the end of this topic you will be able to:

- state that the mains supply is a.c. and a battery supply is d.c.
- explain, in terms of current, the terms a.c. and d.c.
- state that the frequency of the mains supply is 50 Hz
- state that the declared value of the mains voltage is 230 V
- draw and identify the circuit symbols for a battery, fuse, lamp, switch, resistor, capacitor, diode and variable resistor
- state that electrons are free to move in a conductor
- describe the electric current in terms of the movement of charges round a circuit
- use correctly the units ampere and volt
- state that the declared value of an alternating voltage is less than its peak voltage
- carry out calculations involving the relationship between charge, current and time
- use correctly the unit 'coulomb'
- state that the voltage of a supply is a measure of the energy given to the charges in a circuit.

Mains and batteries

Mains electricity is alternating current (a.c.). It changes its direction 100 times every second. The voltage from the mains can be viewed on an oscilloscope.

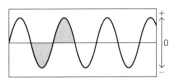

Alternating voltage

In one cycle, the current flows one way for one-hundredth of a second, then the other way for an equal time.

The mains frequency is 50 hertz (50 Hz), which means there are 50 cycles per second, i.e. the current flows one way and then the other 50 times every second.

The declared value of the mains voltage is 230 volts (230 V).

Batteries provide direct current (d.c.). This is because the negatively charged electrons **always** move away from the negative side of the battery and are attracted to the positive side so there is no change in direction. The voltage from a battery can be viewed on an oscilloscope.

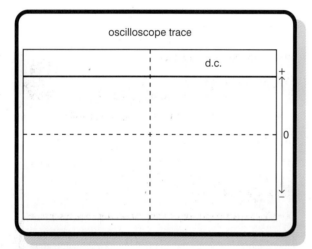

Direct current voltage

Electrons (current) will only flow in one direction, from (−) to (+).

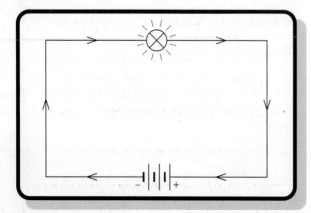

Current flowing from − to +

Electrical symbols

Electrical circuits are made up of various components. Each component has its own symbol, which must be used correctly.

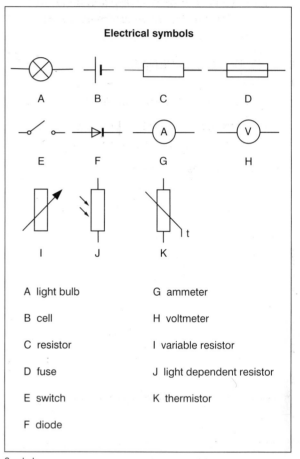

Symbols

Movement of charges

In a conductor, the electrons are free to move around and they do so in a random manner. If a battery is connected across the conductor (for example a wire) then all the electrons will move in one direction through the conductor. An electric current is a flow of electrons (or charge) round a circuit or wire. The greater the flow of charge in a circuit, the greater the current.

A battery or cell voltage (measured in **volts,** V) is a measure of the energy needed to make the electrons move round the circuit.

Using electricity

Conductors and insulators

1 **In an insulator**, all the electrons are firmly attached to atoms.
2 **In a conductor**, such as a metal, some electrons have broken away from the atoms. These electrons move about freely inside the conductor.

Current and charge

Fact file

★ An **electric current** is a flow of electric charge.

★ The unit of electric current is the **ampere** (A); the unit of electric charge is the **coulomb** (C).

★ **Electrons** are responsible for the flow of charge through a **metal** when it conducts electricity.

Steady currents

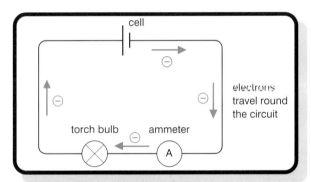

A steady current

The diagram shows a simple electric circuit in which a cell is connected in series with an ammeter and a torch bulb.

1 **The current in this circuit is constant**. This means that charge passes through each component at a steady rate. Electrons pass through each component at a steady rate.
2 **The current through each component is the same**. Equal amounts of charge pass through each component in a set period of time. The number of electrons per second passing through each component is the same.

Questions

1 a) What type of charge, positive or negative, does an electron carry?
b) In a circuit containing a battery, do electrons leave the battery at its positive terminal or its negative terminal?
2 State the unit of a) current b) voltage.

Equation for charge flow

★ **One coulomb of charge** is defined as the amount of charge passing a point in a circuit in one second when the current is one ampere.

★ **The charge passing a point in a circuit** that is carrying a steady current is calculated as follows:

charge passed = current × time taken
(in C)   (in A)   (in s)

$$Q = It$$

This is the QuIt equation.

Worked example

What current flows when 900 C of charge flows round a circuit in a time of 5 minutes?

Solution

$I = \frac{Q}{t} = \frac{900}{5 \times 60} = 3\,\text{A}$

At a junction

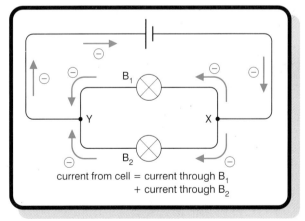

At a junction

Answers
1 a) Negative b) Negative
2 a) Ampere b) Volt

In this circuit, two torch bulbs B_1 and B_2 are in parallel with each other. Each electron from the cell passes through one torch bulb then returns to the positive terminal of the cell. The flow of electrons from the cell divides at junction X and recombines at junction Y.

The charge flow per second from the cell = charge flow per second through B_1 + charge flow per second through B_2. Since charge flow per second is current, it follows that cell current = current through B_1 + current through B_2.

In other words, at a junction in a circuit.

The total current entering the junction is equal to the total current leaving the junction.

Question

3 a) Calculate the charge that flows through a torch bulb carrying a steady current of 0.25 A in
(i) 1 minute (ii) 5 minutes.
b) Calculate how long it takes for a charge of 100 C to pass through a torch bulb that is carrying a steady current of 0.2 A.

Voltage

This is the electrical push which makes electrons move round a circuit. Electrons need energy before they can move. A voltage of 1 V means that each coulomb of charge gets 1 joule of energy.

Peak voltage

Mains voltage is declared as 230 V a.c. but this is only an average value or **root mean square** (rms) as it is changing all the time.

The maximum voltage reached is about 325 V and this is referred to as the **peak value**. This means that the declared value of an a.c. voltage is less than its peak value.

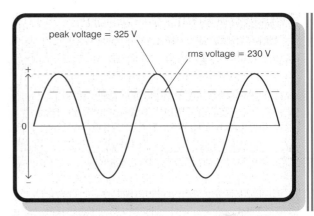

Peak and rms values

2.3 Resistance

preview

At the end of this topic you will be able to:
- draw and identify the circuit symbols for an ammeter and a voltmeter
- draw circuit diagrams to show the correct positions of an ammeter and a voltmeter in a circuit
- state that an increase in the resistance of a circuit leads to a decrease in the current in that circuit
- carry out calculations involving the relationship between resistance, current and voltage
- use correctly the unit ohm
- give two practical uses of variable resistors
- state than when there is an electric current in a wire, there is an energy transformation
- give three examples of resistive circuits in the home in which electrical energy is transformed into heat
- state that the electrical energy transformed each second is VI
- state the relationship between energy and power
- use correctly in context the terms 'energy', 'power', 'joule' and 'watt'

Answer: 3 a) (i) 15 C (ii) 75 C b) 500 s

Using electricity

- carry out calculations involving the relationship between power, current and voltage
- state that in a lamp, electrical energy is transformed into heat and light
- state that the energy transformation in an electric lamp occurs in the resistance wire (filament lamp) or gas (discharge tube)
- state that a discharge tube lamp is more efficient than a filament lamp (i.e. more of the energy is transformed into light and less into heat)
- state that the energy transformation in an electric heater occurs in the resistance wire (element)
- state that **V/I** for a resistor remains approximately constant for different currents
- explain the equivalence of **VI** and **I²R**
- carry out calculations using the relationship between power, current and resistance.

Ammeters and voltmeters

Ammeter symbol   Voltmeter symbol

Ammeters measure current and voltmeters measure voltage. Ammeters must be connected in **series** to measure the current **through** a component.

Voltmeters must be connected in **parallel** to measure the voltage **across** a component.

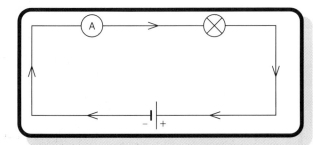
Ammeter in a circuit

To insert an ammeter the circuit has to be broken. To insert a voltmeter the circuit does not have to be altered.

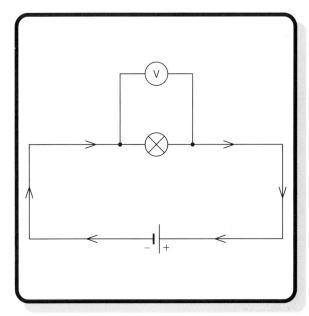
Voltmeter in a circuit

Resistance
Fact file

★ When a component opposes the flow of current in a circuit, it is said to have resistance. If resistance **increases** then the current **decreases**:

if $R \uparrow$ then $I \downarrow$

★ The **resistance** R of an electrical component in a circuit = $\dfrac{\text{the voltage across the component}}{\text{the current through the component}}$

★ The unit of resistance is the **ohm** (Ω), defined as one volt per ampere.

★ **To calculate resistance**, use the equation

resistance = $\dfrac{\text{voltage}}{\text{current}}$

★ A **resistor** is a component designed to have a particular value of resistance. This resistance is caused by opposition to the motion of electrons round the circuit.

Note
The following prefixes are used for large or small values of current, voltage or resistance:

mega (M)	kilo (k)	milli (m)	micro (μ)
1 000 000	1000	0.001	0.000 001
10^6	10^3	10^{-3}	10^{-6}

Resistance, current and voltage

The formula which connects voltage, current and resistance is

voltage = current × resistance

$$V = IR$$

Worked example
What is the resistance of a component when the voltage across it is 12 V and the current through it is 0.25 A?

Solution
$R = V/I = 12/0.25 = 48\,\Omega$

What does current depend on?
1. the voltage of the cell, battery or power supply unit
2. the resistance of the components in the circuit.

Resistors can be used to control the current in a circuit.

Variable resistors

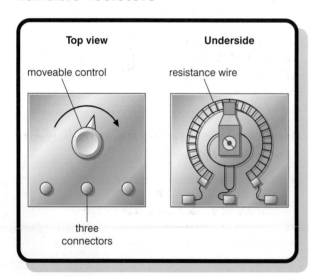

A variable resistor

The resistance of a variable resistor is normally altered by changing the length of wire inside it, i.e. the shorter the length of wire, the less resistance it has.

Variable resistors are used as volume controls on radios, heating controls and as lamp dimmers.

Energy transformation
As electrons push their way through a resistor, they transfer energy in the form of heat. Some household appliances use this energy transformation to produce heat in a resistance wire or element, for example electric fires, toasters and cookers.

Power
The rate at which electrical energy is charged into other forms of energy is equal to electrical power. Power is measured in watts (W) where 1 watt is equal to 1 joule per second (1 W = 1 J/s).

$$\text{power (W)} = \frac{\text{energy (J)}}{\text{time (s)}} \quad \text{and}$$

power = voltage × current
(W)   (V)   (A)

$$P = \frac{E}{t} \qquad P = VI$$

Worked example
Find the power of an appliance which changes 12 000 J of electrical energy into heat in 10 minutes.

Solution
$P = E/t = 12\,000/(10 \times 60) = 20\,\text{W}$

Worked example
A mains appliance uses 230 V and has a power rating of 920 W. Find the current in the appliance.

Solution
$I = P/V = 920/230 = 4\,\text{A}$

Let there be light
Electric lights can be divided into two distinct types: filament lamps and fluorescent lamps.

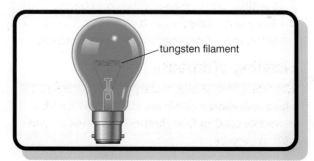

A filament lamp

The filament is made of tungsten wire, which becomes very hot and gives out white light. Only about 10% of the electrical energy is given out as light energy, the rest is wasted as heat.

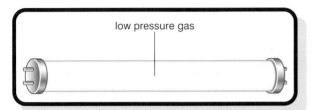

A fluorescent lamp

A high voltage across a low pressure gas inside the glass tube produces rays, which release white light when they hit the coating inside the tube. About 40% of the electrical energy supplied is transformed into light energy. A fluorescent lamp is thus more efficient than a filament lamp.

Resistance heating

All the electrical energy supplied to a resistor is transformed to heat energy, heating the surroundings as a result.

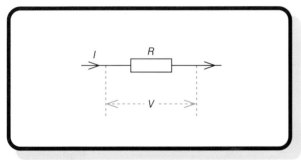

Heating effect

1. The voltage V across the resistor $= IR$, where I is the current through the component.
2. The power supplied to the resistor $P = IV$. This is the rate of heat produced in the resistor.

Heating elements

Elements have a high resistance and convert electrical energy into heat energy. The heating elements used in fires, kettles and irons all use nichrome wire.

Questions

1 A steady current of 3 A is passed through a 5 Ω resistor. Calculate
   a) the charge that passes through this resistor in 1 minute
   b) the voltage across the resistor.

2 When a current of 2 A was passed through a certain resistor, the voltage across the resistor was 10 V. Calculate
   a) the resistance of this resistor
   b) the power supplied to this resistor.

Ohm's law

Ohm discovered by experiment that if a conductor was kept at a constant temperature, then the **voltage divided by the current** was constant, i.e. V/I = a constant.

He found that if the voltage doubled, the current doubled as well. If the voltage trebled, so did the current.

The graph of voltage against current is a straight line through the origin. This means that the current through a resistor is directly related to the voltage across it or V/I = a constant.

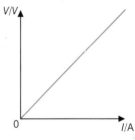

Ohm called the constant the **resistance** (R) of the resistor, measured in ohms.

> Ohm's law, V/I ($=R$), is constant for a conductor at constant temperature.

Answers
1 a) 180 C b) 15 V
2 a) 5 Ω b) 20 W

Measuring resistance

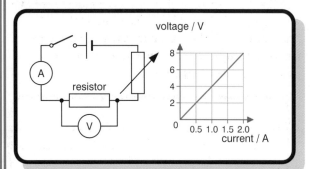

Measuring resistance   Voltage against current

The diagram shows how the resistance of a resistor may be determined using an ammeter and a voltmeter. Note that the voltmeter is in parallel with the resistor, and the ammeter is in series.

1. With the switch closed, the variable resistor is adjusted to change the current in steps. At each step, the current and voltage are measured from the ammeter and the voltmeter respectively.
2. The measurements are plotted as a graph of voltage (on the vertical axis) against current, as shown. The plotted points define a straight line passing through the origin.
3. Read off the voltage and current for a point on the line near the top. The voltage divided by the current is equal to the resistance of the resistor.

Question

3 a) Determine the resistance of the resistor that gave the results plotted on the graph.
 b) Calculate the current through this resistor when the voltage across it is (i) 5.0 V (ii) 0.1 V.

More about power

Fact file

★ An **electric current** is a **flow of charge**. The unit of electric current is the **ampere** (A).

★ The **potential difference** (p.d.) between any two points in a circuit is the **work done per unit charge** when charge moves from one point to the other point. Voltage is an alternative word for potential difference.

★ The unit of charge is the **coulomb** (C). One coulomb is equal to the charge passing a point in a circuit in one second when the current is one ampere.

★ The unit of potential difference is the **volt** (V), equal to one joule per coulomb.

Consider a component in a circuit which has voltage V across its terminals and which passes a steady current I.

1. In a time interval t, charge Q flows through the component where $Q = It$.
2. The electrical energy E delivered by charge Q is QV, since V is defined as the electrical energy delivered per unit charge.
3. Hence the electrical energy E delivered in time t is ItV.
4. The electrical power = $\dfrac{\text{electrical energy delivered}}{\text{time taken}} = IV$.

Power, current and resistance

Ohm's law, $V = IR$, and the equation for power, $P = VI$, can be combined to give a formula for calculating power using only current (I) and resistance (R).

$V = IR$ (1)

$P = IV$ (2)

substitute $V = IR$ into equation (2)

$P = (IR)I = I^2R$

$P = I^2R$

$\boxed{P = I^2R}$

Worked example

Calculate the current flowing in a 2 kW heater if the resistance of the element is 200 Ω.

Answer

3 a) 4.0 Ω   b) (i) 1.25 A   (ii) 25 mA

Solution

$I^2 = P/R = 2000/200 = 10$

$I = \sqrt{10}$

$= 3.17\,A$

Worked example

Calculate the resistance of an element which has a power rating of 3000 W and uses a current of 4 A.

Solution

$R = P/I^2 = 3000/4^2 = 3000/16 = 187.5\,\Omega$

2.4 Useful circuits

preview

At the end of this topic you will be able to:

- state a practical application in the home which requires two or more switches used in series
- state that in a series circuit the current is the same at all points
- state that the sum of the currents in parallel branches is equal to the current drawn from the supply
- explain that connecting too many appliances to one socket is dangerous because a large current could be drawn from the supply
- state that the voltage across components in parallel is the same for each component
- state that the sum of voltages across components in series is equal to the voltage of the supply
- describe how to make a simple continuity tester
- describe how a continuity tester may be used for fault finding
- draw circuit diagrams to describe how the various car lighting requirements are achieved
- carry out calculations involving the relationships $R_t = R_1 + R_2 + \ldots$
  $1/R_T + 1/R_1 + 1/R_2 + \ldots$

Using electricity

Series switches

Mains appliances such as washing machines have to be switched on at the socket and at the machine itself. This is done using two switches in series. Unless both switches are closed the machine will not operate.

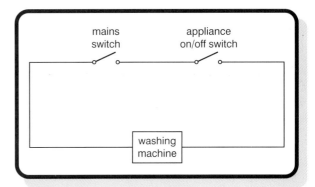

Two switches in series

Series circuits

A series circuit is one which provides the current with only one path, i.e. the current flows round the circuit without splitting up.

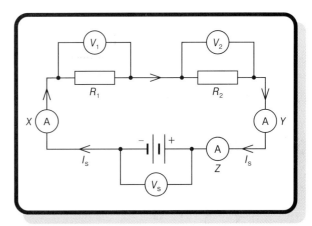

Series circuit

In a series circuit the current is the same at every point. This can be verified by placing ammeters at various points in the circuit, e.g.

$I_X = I_Y = I_Z$

The sum of the voltages (measured with a voltmeter) across components in a series circuit is equal to the voltage of the supply:

$V_1 + V_2 = V_S$

Parallel circuits

A parallel circuit is one which provides more than one path for the current, i.e. the current splits up as it moves round the circuit.

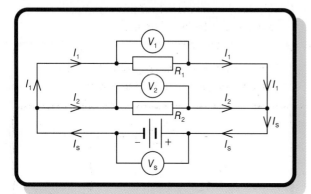

Parallel circuit

The same current arrives back at the battery, i.e. I_s leaves and I_s returns. The current splits through parallel branches and then recombines. No electrons are lost. This can be stated in the equation
$I_s = I_1 + I_2$

The voltage across each parallel component is the same as the supply voltage V_s, i.e.
$V_s = V_1 = V_2$

$R_T = \dfrac{R_1 \times R_2}{R_1 + R_2}$

Danger

Connecting too many appliances to one socket in the home is dangerous because a large current could be taken from the mains supply, leading to overheating in the flexes which could in extreme cases start a fire.

Continuity tester

A continuity tester can be used to find faults in wires. A simply continuity tester is shown below where points P and Q are connected to a suspect wire.

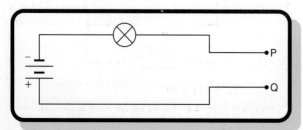

Continuity tester with battery and bulb

If the points P and Q are connected to the ends of a wire under test, then if the wire is not fault, the bulb will light. If the wire is broken inside the plastic insulation, then the bulb will not light. A broken wire is said to be an **open circuit** wire.

An ohmmeter can be used to detect an open circuit wire. If the wire is not faulty then the ohmmeter reads $0\,\Omega$, but if the wire is open circuit, i.e. broken, the ohmmeter shows an extremely large value of resistance.

An ohmmeter can also be used to detect faults in bulbs. A bulb can be in three conditions:

1. If the bulb is not faulty, then the ohmmeter will read a few ohms, which is the resistance of the filament.
2. If the filament of the bulb is broken, i.e. it is an **open circuit** bulb, then the ohmmeter shows an extremely large value.
3. If the wires at the base of the bulb are touching, which would cause the bulb not to light, then the bulb is said to be a **short circuit** bulb and the ohmmeter reads $0\,\Omega$.

Question

1. Complete **a)** and **b)** by inserting the correct word in each blank.
   In a circuit in which two components are connected together,
   a) the current is always the same if the two components are connected in _____ .
   b) the voltage is always the same if the two components are connected in _____ .

Answer 1 a) Series b) Parallel

Using electricity

Car electrics

All lights on a car are connected in **parallel** so they each get 12 V from the battery.

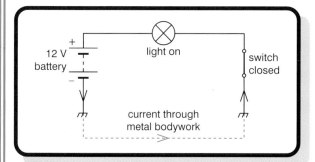

Sidelight

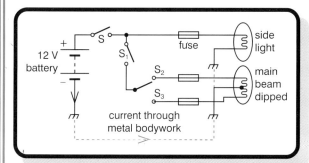

Sidelight and headlight

Note: switches S_2 and S_3 are as shown to prevent the main and dipped beams being on at the same time.

Resistors in series and parallel

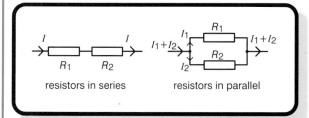

Resistance rules

For two or more resistors or resistance R_1, R_2, etc. in series,

total resistance $R_T = R_1 + R_2$ + etc.

For two resistors of resistance R_1 and R_2 in parallel, the total resistance R_T is given by the equation

$\frac{1}{R_T} = \frac{1}{R_1} + \frac{1}{R_2}$

Notes

a) *When dealing with resistors in parallel, the total resistance R_T is always smaller than the smallest resistor value present.*

b) *The total resistance R_T of two resistors in parallel can be found using:*

$R_T = R_1 R_2 / (R_1 + R_2)$

*This can be remembered as 'product over sum' or the '**MAD**' rule (**m**ultiply, **a**dd then **d**ivide).*

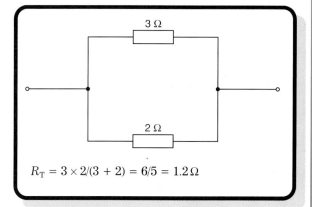

Two resistors in parallel

c) The total resistance R_T of two resistors in parallel of the same value is equal to half the value of one resistor, e.g. a $10\,\Omega$ and a $10\,\Omega$ in parallel is equal to $5\,\Omega$. A $3\,\Omega$ and a $3\,\Omega$ in parallel is equivalent to $1.5\,\Omega$.

d) For any number (n) of resistors in parallel, all of the same value, the total resistance R_T is equal to R/n where R is the value of any one resistor.

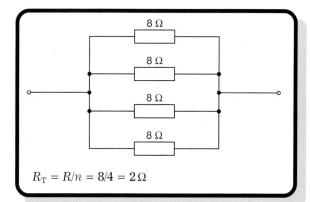

Four resistors in parallel

Worked example

Find the current in the following circuit and the voltage across each resistor.

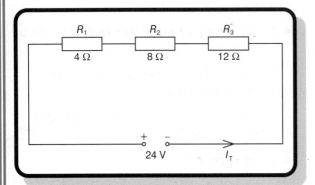

Resistors in series with 24 V supply

Solution

$R_T = R_1 + R_2 + R_3$

$R_T = 4 + 8 + 12 = 24\,\Omega$

$I_T = V_S/R_T = 24/24 = 1\,A$

$V_{R1} = I_T \times R_1 = 4\,V$

$V_{R2} = I_T \times R_2 = 8\,V$

$V_{R3} = I_T \times R_3 = 12\,V$

Worked example

Find I_T, I_1 and I_2 in the circuit below.

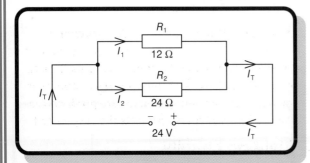

Resistors in parellel with 24 V supply

Solution

$R_T = \text{product/sum} = (12 \times 24)/(12 + 24) = 8\,\Omega$

$I_T = V_S/R_T = 24/8 = 3\,A$

$I_1 = V_S/R_1 = 24/12 = 2\,A$

$I_2 = V_S/R_2 = 24/24 = 1\,A$

Note: $I_T = I_1 + I_2$

Questions

2 A $3\,\Omega$ resistor, a $6\,\Omega$ resistor and a 9 V battery are connected in series. Calculate **a)** the total resistance of the two resistors
b) the current supplied by the battery.

3 A $12\,\Omega$ resistor and a $4\,\Omega$ resistor are connected in parallel to the terminals of a 3 V battery. Calculate
a) the total resistance of the two resistors
b) the current supplied by the battery
c) the current passing through each resistor.

Answers

2 a) $9\,\Omega$ **b)** 1 A
3 a) $3\,\Omega$ **b)** 1 A
c) 0.25 A through the $12\,\Omega$ resistor, 0.75 A through the $4\,\Omega$ resistor.

2.5 Behind the wall

preview

At the end of this topic you will be able to:
- state that household wiring connects appliances in parallel
- state that mains fuses protect the mains wiring
- state that a circuit breaker is an automatic switch that can be used instead of a fuse
- state that kWh is a unit of energy
- describe, using a circuit diagram, a ring circuit
- state the advantages of using the ring circuit as a preferred method of wiring in parallel
- give two differences between a lighting circuit and a power ring circuit
- state one reason why a circuit breaker may be used in preference to a fuse
- explain the relationship between kilowatt hours and joules.

Alternating current

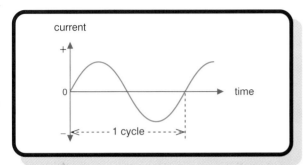

Alternating current

1. The electric current through a mains appliance alternates in direction. The current reverses direction then reverses back each cycle.
2. The **frequency** of an alternating current is the number of cycles per second. In the UK, the mains frequency is 50 Hz.

Household wiring

The ring main circuit consists of a closed loop of three-core cable (live, neutral and earth) which connects the sockets in a house. All the sockets are connected in parallel, so appliances (when in use) will get 230 V from the mains without having any effect on other appliances.

Mains fuses

Every house has a fuse box or consumer unit similar to that shown below.

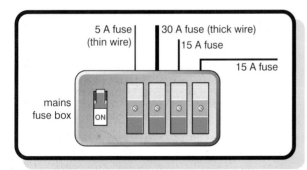

Mains fuse box

This will contain either fuses or circuit breakers, which are designed to protect the mains wiring. More commonly nowadays consumer units contain circuit breakers, which are automatic cut-off switches that do the same job as fuses.

Electricity costs

★ One kilowatt hour (kWh) is the electrical energy supplied to a one kilowatt appliance in exactly one hour.

★ The kilowatt hour is the unit of electricity for costing purposes.

★ A domestic electricity meter records the total number of units used.

★ Number of units (kWh) = power rating of appliance (kW) × time used (h).

A ring circuit

This is a special parallel circuit used in houses to connect all the power sockets together. Doing this enables each socket to operate at 230 V.

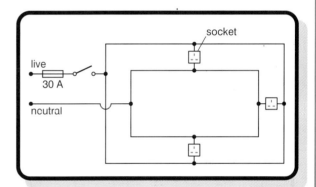

Simplified ring circuit diagram

A ring circuit has several advantages over a simple parallel circuit. Current can reach any socket by two routes. To supply 30 A, each route carries only 15 A, so **thinner, cheaper cables** can be used. Consequently, **less heat** is produced in the cables. A 30 A fuse protects the ring circuit.

Lighting circuits

The lighting circuit in a house is **not** a ring circuit. A single cable joins all the ceiling lights in **parallel**. The current used by all the lights does not exceed 5 A. Thinner cable can be used and there is no need for a ring circuit. The circuit is protected by a 5 A fuse.

Circuit breakers

Circuit breakers protect circuits from current overload. They replace fuses because they switch off or trip as soon as a slight overload happens. They are easily reset by pressing a button and are tamper-proof.

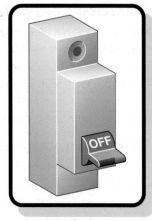

Circuit breaker

Energy in a unit (1 kWh)

There are 3 600 000 J of energy in a kilowatt hour. This is calculated using the equation:

$$E\,(J) = P\,(W) \times t(s)$$
$$= 1\,kW \times 1\,hour$$
$$= 1000\,W \times 3600\,s$$
$$= 3\,600\,000\,J$$

Questions

1. State one advantage and one disadvantage of a circuit breaker in comparison with a fuse.
2. Why is the wiring for a ring main circuit thicker than the wiring of a lighting circuit?

Mains circuits

Each circuit from the fuse board is protected with its own fuse. If the fuse 'blows', the live wire is therefore cut off from appliances supplied by that circuit.

The mains cable from the substation to a building is connected via the electricity meter to the circuits in the building at the distribution fuse board. The live wire from the substation is connected via a main fuse to the electricity meter.

The two wires used to supply an electric current to an appliance are referred to as the **live** and the **neutral** wires. The neutral wire is earthed at the nearest mains substation.

Mains wires need to have as low a resistance as possible, otherwise heat is produced in them by the current. This is why mains wires are made from copper. All mains wires and fittings are insulated.

The fuse in a lighting circuit is in the fuse box. Each light bulb is turned on or off by its own switch. When the switch is in the off position, the appliance is not connected to the live wire of the mains supply.

A **ring main** is used to supply electricity to appliances via wall sockets. A ring main circuit consists of a live wire, a neutral wire and the **earth wire** which is earthed at the fuse board. The wires of a ring main are thicker than the wires of a lighting circuit because appliances connected to a ring main require more current than light bulbs do.

Each appliance is connected to the ring main by means of a three-pin plug which carries a fuse. An appliance with a metal chassis is earthed via the three-pin plug and the earth wire. This prevents the metal chassis from becoming live if a fault develops in the appliance. Appliances connected to the ring main can be switched on or off independently since they are in parallel with each other.

Mains circuits

Answers

1. *Advantage:* a circuit breaker does not need to be replaced each time it trips; *disadvantage:* a circuit breaker is more expensive than a fuse.
2. Ring main currents are greater so the wires need to have less resistance otherwise they would become warm when current passes through them.

Using electricity

2.6 Movement from electricity

preview

At the end of this topic you will be able to:
- **identify on a simple diagram of an electric motor the rotating coil, field coil (magnet), brushes and commutator**
- **state that a magnetic field exists around a current-carrying wire**
- **give two examples of practical applications which make use of the magnetic effect of a current**
- **state that a current-carrying wire experiences a force when the wire is in a magnetic field**
- **state that the direction of a force on a current-carrying wire depends on the direction of the current and of the field**
- **explain the operation of a d.c. electric motor in terms of forces acting on the coil and the purpose of the brushes and commutator**
- **state the reasons for the use in commercial motors of carbon brushes, multi-section commutators and field coils.**

Magnetism

Fact file

★ Two magnets act on each other at a distance. The force of interaction decreases with increased distance.

★ The law of force for magnetic poles is that **like poles repel and unlike poles attract**.

★ Iron and its steel alloys can be magnetised and demagnetised.

★ Permanent magnets are made from steel because steel is hard to demagnetise once magnetised.

★ Electromagnets are made from soft iron because it is easy to magnetise and demagnetise.

Magnetic field patterns

The lines of force of a magnetic field are defined by the direction of a plotting compass in the magnetic field. A bar magnet suspended on a thread aligns itself with the Earth's magnetic field. The end that points north is the north-seeking pole; the other end is the south-seeking pole.

Permanent magnets

A bar magnet produces lines of force which loop round from one end to the other end. The lines emerge from the north-seeking pole and end at the south-seeking pole.

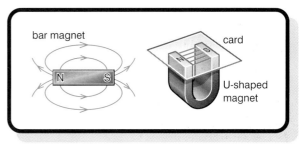

Magnetic fields

The magnetic effect of a steady electric current

When a wire has a current flowing through it, a magnetic field is produced around the wire.

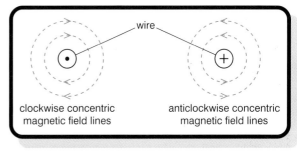

(a) Current out of paper   (b) Current into paper

An electromagnet is a current-carrying wire wrapped around an iron core. Powerful electromagnets use large currents and many turns of wire in the coil.

Electromagnets

An electromagnet consists of a solenoid with an iron core. When a current is passed through the coil of wire, the iron bar is magnetised. The magnetic field is much stronger than the field created by the empty coil. It can be switched off by switching the current off, and its strength can be altered by changing the current.

Using electromagnets

Because they can be switched off or on as required, unlike a permanent magnet, electromagnets have many applications. Some examples are bells, door chimes, relays, loudspeakers and motors, and they are even used in hospitals to remove metal splinters from eyes.

To lift scrap iron: a powerful electromagnet suspended from a crane cable is used to lift scrap iron in a scrapyard. Switching the electromagnet off causes the scrap iron to fall off the electromagnet.

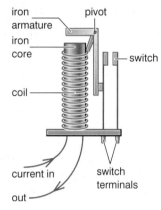

A normally open relay

The relay: when current is passed through the coil of a relay, the electromagnet attracts a soft iron armature. The movement of this armature opens or closes a switch which is part of a different circuit. When the current is switched off, the armature springs back to its normal position and the switch reverts to its original state.

The electric bell: the electromagnet coil is part of a 'make-and-break' switch. When current passes through the coil, the electromagnet attracts the soft iron armature which makes the hammer hit the bell. The movement of the armature opens the 'make-and-break' switch which switches the electromagnet off, allowing the armature to spring back to its initial position and close the switch. Current then passes through the electromagnet again, causing the sequence to be repeated.

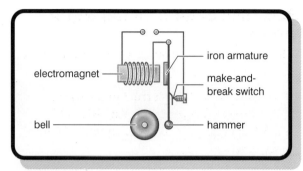

An electric bell

Questions

1. Why is iron better than steel for use in an electromagnet?
2. Why is steel better than iron for use as a permanent magnet?

The motor effect

A force is exerted on a current-carrying wire placed at right angles to the lines of force of a magnetic field. The diagram shows a current-carrying wire between the poles of a U-shaped magnet. The force is perpendicular to the wire and to the lines of force of the magnet.

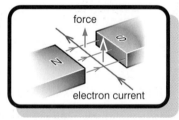

The motor effect

The force occurs because the magnetic field created by the wire interacts with the applied magnetic field (the magnetic field in which the wire is placed). The combined field is very weak on one side where the two fields are in opposite directions to each other. The force acts towards the side where the combined field is very weak.

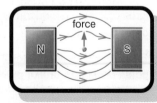

Combined fields

Answers

1. Iron loses its magnetism when the current is switched off whereas steel keeps its magnetism.
2. Steel keeps its magnetism whereas iron loses it.

Using electricity

Direction of force in the motor effect

The direction of force on a current-carrying wire in a magnetic field depends on two factors:

1. the direction of the current in the wire
2. the direction of the other magnetic field.

If **1** or **2** above is reversed, the direction of force is reversed.

The operation of a d.c. motor

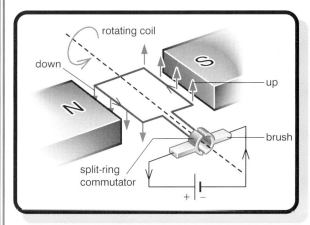

The d.c motor

The three main parts of a simple d.c. motor are:

1. the permanent magnets, which have a magnetic field between their poles
2. the rotating coil
3. the commutator and brushes.

When a current flows into the coil (as shown above), forces act on opposite sides of the coil to make it turn anticlockwise. A metal cylinder which is split into two halves is attached to the ends of the coil and rotates with it. This is called the commutator. The brushes press against the commutator allowing the current to enter the coil from the d.c. supply. The brushes do not rotate. The commutator and brushes reverse the current direction every half turn to ensure that the coil keeps rotating anticlockwise.

Question

3 For the electric motor above, what would be the effect of
  a) reversing the current direction **b)** reversing the current direction *and* reversing the direction of the magnetic field?

Commercial motors

Real or commercial motors, such as those used in vacuum cleaners or washing machines, have carbon brushes, multi-section commutators (instead of just two halves) and field coils.

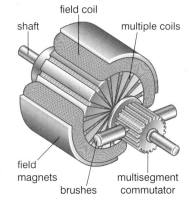

A practical motor

Carbon brushes: carbon (graphite) is a good conductor which can slide easily over a commutator without damaging it. Good electrical contact is ensured using springs at the back of the brushes to press them against the commutator.

Multi-section commutator: a single coil is replaced by several coils spaced around the motor. The commutator is divided into several segments or sections with a coil attached to each pair of segments. This makes the rotation of the motor smoother.

Field coils: permanent magnets are replaced by electromagnets called **field coils**. Electromagnets are also cheaper, lighter, do not lose their magnetism and can be switched off and on when required.

A mains electric motor works with alternating current because it has an electromagnet, not a permanent magnet. Each time the current reverses, the magnetic field does too, so the rotation direction is unchanged.

Answer

3 The rotation direction would **a)** reverse **b)** be unchanged

Revise Standard Grade Physics

Equations review

1 $R = V/I$

This is true for a whole circuit when V is the supply voltage, I is the current sent out from the supply and R is the total resistance of the whole circuit, **and also** for any single component when V is the voltage across it, I is the current through it and R is its resistance

2 $R_T = R_1 + R_2$

Resistors in series have a total resistance equal to the sum of the separate resistances

3 $R_T = \dfrac{R_1 \times R_2}{(R_1 + R_2)}$

or

$\dfrac{1}{R_T} = \dfrac{1}{R_1} + \dfrac{1}{R_2}$

$R_T = R/n$

Resistors in parallel have a total resistance less than any of the separate resistances. The first equation (the MAD rule) only works for two resistors in parallel. MAD means **m**ultiply, **a**dd then **d**ivide. This is true for n resistors of the same value in parallel

4 $Q = It$

The QuIt equation: Q is the electrical charge in coulombs (C), t is the time in seconds and I is the current in amps (A)

5 $P = E/t$

P here is the power as it is commonly used, that is the rate at which energy E is transferred. It is measured in joules per second (J/s) or watts (W)

6 $P = VI$

We have already met $V/I (=R)$, here is $V \times I$. Often we have to use both of these equations in the same problem

7 $P = I^2R$

This is often used in calculating the rate at which electrical energy is turned into heat in cables or resistors. It is sometimes an alternative to using the two equations above

round-up

1 a) What type of flex is suitable for
  (i) an electric iron
  (ii) a table lamp?
b) What is the colour of each of the three wires of a mains flex? [7]

2 a) What is the difference in terms of electron flow between a direct current and an alternating current?
b) State one advantage of a circuit breaker in comparison with a fuse.
c) Why must a mains fuse or a circuit breaker always be in the live wire? [5]

3 A torch bulb is connected in series with a switch, an ammeter and a 3.0 V battery.
a) Sketch the circuit diagram, showing the switch open
b) When the switch is closed, the ammeter shows a constant reading of 1.5 A. Calculate
  (i) the charge that passes through the torch bulb each minute
  (ii) the energy delivered to the torch bulb each minute. [4]

4 a) Sketch a circuit showing a 10 Ω resistor in series with a 5 Ω resistor and a 1.5 V cell.
b) (i) Calculate the total resistance of these two resistors in series.
  (ii) Calculate the current through each resistor in the circuit. [3]

5 In the circuit below, the ammeter reading was 0.25 A and the voltmeter reading was 3.0 V.

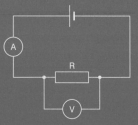

a) Calculate the resistance of the resistor R. [1]
b) If a second resistor identical to R was connected in parallel with R, how would the voltmeter and ammeter readings alter? [2]

Using electricity

6 a) How many different resistance values can be obtained using a 2 Ω, a 3 Ω and a 6 Ω resistor? Sketch each possible combination.

b) Calculate the maximum and minimum total resistance in this set of combinations [8]

7 The diagram below shows a 3 Ω resistor in parallel with a 6 Ω resistor. The two resistors are connected across the terminals of a 1.5 V cell.

a) Calculate the current through each resistor and the current through the cell.

b) Calculate the total resistance of these two resistors in parallel. [4]

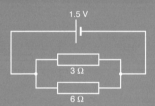

8 In the circuit shown below ammeter A_1 reads 0.5 A and ammeter A_2 reads 0.3 A

a) Calculate the current through resistor Y. [1]

b) The voltmeter in parallel with resistor Y reads 1.0 V when ammeter A_1 reads 0.5 A. Calculate the resistance of resistor Y. [2]

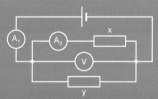

9 a) A 3.0 V battery and two 10 Ω resistors are connected in series. Draw a circuit diagram and calculate the current and the voltage for each resistor.

b) A third 10 Ω resistor is connected in parallel with one of the resistors in the above circuit. Draw the new circuit diagram and calculate the current and the voltage for each resistor in this new circuit. [11]

10 a) Describe the energy changes that take place in a filament lamp.

b) Which is more efficient, a discharge tube lamp or a filament lamp? [4]

11 A current of 2.5 A is passed through a 6.0 Ω resistor. Calculate **a)** the voltage across the resistor

b) the power supplied to the resistor. [2]

12 An electric heater is rated at 230 V, 1000 W. Calculate

a) the electrical energy delivered to the heater in 300 s

b) the current through the heater when it operates at its rated power. [3]

13 The circuit diagram below shows two 5 Ω resistors P and Q in series with each other, an ammeter and a 3.0 V cell.

a) (i) What is the voltage across each resistor? [1]
(ii) What is the current through each resistor? [1]

b) Hence calculate the power supplied to each resistor. [2]

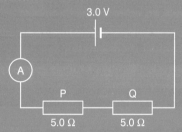

c) A third 5 Ω resistor R is then connected in parallel with resistor P. How does this affect the current passing through each of the other two resistors? [2]

14 a) The diagram shows a vertical wire between the poles of a U-shaped magnet, arranged so that the magnetic field of the magnet is horizontal.

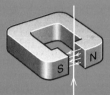

When a current passes up the wire, the wire is forced inwards. If the magnetic field is reversed and the current is reversed, in which direction will the force on the wire then be? [1]

b) What would happen to the wire in **a)** if an alternating current were used? [2]

Total = 66 marks

Health physics

3.1 The use of thermometers

MIND MAP Page 9.

preview

At the end of this topic you will be able to:
- state that a thermometer requires some measurable physical property that changes with temperature
- describe the operation of a liquid in glass thermometer
- describe the main differences between a clinical and an ordinary thermometer
- describe how body temperature is measured using a clinical thermometer
- explain the significance of body temperature in diagnosis of illness.

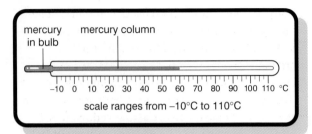

A mercury thermometer

Ordinary and clinical thermometers

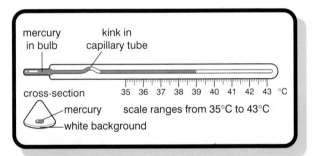

A clinical thermometer

What thermometers have in common

There are many different kinds of thermometers and they come in a variety of shapes and sizes. However, they all have one thing in common: a physical property which changes with temperature and which can be measured. Two examples are:

1. in a liquid crystal thermometer, tiny crystals have different colours at different temperatures
2. in a platinum resistance thermometer, its resistance changes with temperature.

A liquid in glass thermometer

Mercury and alcohol thermometers contain a liquid which expands as it gets hotter and contracts as it gets cooler. This means that the volume of the mercury or alcohol changes and therefore the length of the mercury or alcohol column in the capillary tube changes.

Clinical thermometers are different from ordinary thermometers in a number of ways. They have:

1. a very narrow tube so that even small temperature changes will still cause the mercury column to move by a reasonable amount
2. a small temperature range, from about 35 °C to 43 °C
3. a kink to prevent the mercury contracting back down the tube and changing the reading when the thermometer is removed from the patient
4. a triangular shape to magnify the narrow mercury column and make the reading of the temperature easier.

How body temperature is measured

A clinical thermometer is used to measure body temperature by firstly shaking it to make the mercury go back into the bulb and then placing it under the tongue. It is left for 1 or 2 minutes until it reaches body temperature and then removed so that a reading can be taken.

Body temperature and diagnosis of illness

Body temperature can be used as the first stage of diagnosis of illness. Normal body temperature is 37 °C and is the temperature of the internal organs of the body. If body temperature falls below 35 °C or rises above 40 °C, symptoms of illness appear. For example, a body temperature of 34 °C will cause shivering and reduce heart rate. Temperatures around 28 °C will result in death. A body temperature of about 41 °C will cause fever and an increased heart rate. Above 43 °C, death will occur.

Questions

1. What is the temperature of each of the two thermometers shown on page 50?
2. a) What is the body temperature of a normal healthy person?
   b) Why is there a kink in the tube of a clinical thermometer?

Answers

1 60 °C, 39 °C
2 a) 37 °C b) To prevent the mercury from returning to the bulb after the thermometer is removed from the patient when the reading is made.

3.2 Using sound

preview

At the end of this topic you will be able to:

- state that a solid, liquid or a gas is required for the transmission of sound
- explain the basic principles of the stethoscope as a 'hearing aid'
- give an example of the use of ultrasound in medicine, e.g. images of an unborn baby
- state that high frequency vibrations beyond the range of human hearing are called ultrasounds
- give two examples of noise pollution
- give examples of sound levels in the range 0 – 120 dB
- state that excessive noise can damage hearing
- explain one use of ultrasound in medicine.

What sound can travel through

Sounds are produced by vibrations and travel as waves through the three states of matter, i.e. solid, liquid and gas.

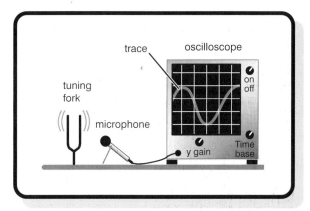

Displaying sound

In a vacuum, there are no solids, liquids or gases to vibrate and so sound cannot travel through a vacuum. This fact is demonstrated by removing the air surrounding a ringing bell in a bell jar. The ringing bell can be heard before the air is pumped out. The sound of the bell fades away as the air is removed. If the air is allowed back into the bell jar, the sound returns.

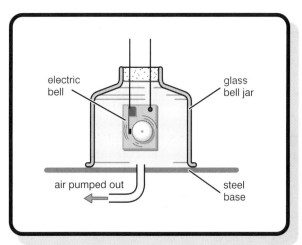

A soundless bell

The stethoscope

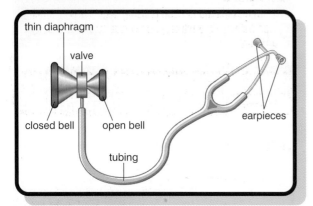

Stethoscope

Doctors listen to sounds from inside the body using a stethoscope. This can help them decide if a patient is ill. Sounds from the heart and lungs are particularly important. A large closed bell with a diaphragm is used to listen to high frequency lung sounds and a small open bell is used to listen to low frequency heart sounds. The sound travels through the air in the stethoscope tubes to the earpieces, which should be tight fitting to keep out external noises and minimise sound losses. The stethoscope tubes should not be too long as the amplitude of the sound decreases considerably between the bell and the earpiece.

Ultrasound

Humans can hear sound in the frequency range 20 to 20 000 Hz. Frequencies above 20 000 Hz cannot be heard by humans although many animals such as bats, dogs and dolphins can hear them. High frequency sounds above 20 000 Hz are called ultrasound.

Ultrasound in medicine

Ultrasound can be used in medicine to produce scans of inside the body. Since it is safe and presents no danger to the cells of the body, it is often used to produce images of unborn babies in the womb of the mother.

Ultrasound can also be used to examine teeth to identify those needing filling. Bursts of ultrasound can be used to break up kidney stones and this removes the need for surgery.

Question

1 a) What is the highest frequency of sound the normal human ear can detect?
b) Explain why sounds produced inside the body that cannot be heard outside the body can be heard using a stethoscope.

Sound levels

The loudness of sound is measured in units called **decibels** (dB) using an instrument called a sound level meter. Some examples are:

normal conversation	60 dB
heavy lorry	100 dB
jet plane	120 dB

Exposure to sound levels above 90 dB over long periods of time can cause loss of hearing. Ear protectors which contain materials that absorb sound should be worn to prevent permanent damage to the ears.

Noise pollution

Very loud sounds can be a health hazard because they may damage hearing. This may arise from any form of noise pollution, such as listening to a personal stereo with the volume turned up too high, working near aircraft or operating noisy machinery.

The graph shows how the lower threshold of hearing varies with frequency.

★ The human ear is most sensitive at a frequency of about 3000 Hz.

★ The upper frequency limit for the human ear is about 20 000 Hz.

★ Hearing usually deteriorates with age – your upper frequency limit will decrease as you become older.

Answer

1 a) 20 000 Hz
b) Sound waves from the body are detected by the diaphragm of the stethoscope and then channelled along the tube to the ear. Without the stethoscope, sound waves from the body spread out so they are much weaker when they reach the ear.

Health physics

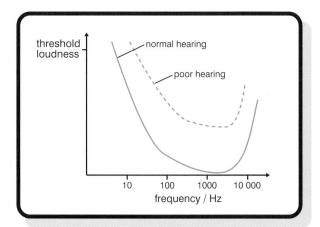

Hearing response

Ultrasound scanners

Ultrasonic waves are produced by a hand-held device called a transducer. This is pressed against the skin over the area being scanned, e.g. the liver or womb. To ensure good contact with the skin and to help with the transmission of sound the area of skin is smeared with a gel. The transducer sends out high frequency ultrasounds of over 1 million hertz (1 MHz) in a narrow beam.

Internal boundaries partially reflect ultrasonic waves, and the reflected waves are detected by the transducer, which is connected to a computer. The transducer builds up an image of the reflecting boundaries inside the body. Unlike X-rays, ultrasonic waves do not cause ionisation, and are therefore thought to be harmless at low power.

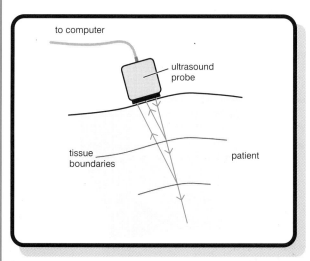

Ultrasound scanner

Question

2 When an ultrasound scan is being taken, why is a gel used between the ultrasound transducer and the body surface?

Answer 2 To ensure the ultrasonic waves enter the body, which they would not do if there was an air gap.

3.3 Light and sight

preview

At the end of this topic you will be able to:

- describe the focusing of light on the retina of the eye
- state what is meant by refraction of light
- draw diagrams to show the change in direction as light passes from air to glass and glass to air
- describe the shapes of convex and concave lenses
- describe the effect of various lens shapes on rays of light
- state that the image formed on the retina of the eye is upside down and laterally inverted
- explain, using a ray diagram, how an inverted image can be formed on the retina
- describe a simple experiment to find the focal length of a spherical convex lens
- state the meaning of long and short sight
- state that long and short sight can be corrected using lenses
- state that fibre optics can be used as a transmission system for 'cold light'
- use correctly in context the terms 'angle of incidence', 'angle of refraction' and 'normal'
- explain, using a ray diagram, how the lens of the eye forms, on the retina, an image of an object (a) some distance from the eye and (b) close to the eye

Revise Standard Grade Physics

- carry out calculations on power/focal length to find either one given the other
- explain the use of lenses to correct long and short sight
- explain the use of fibre optics in an endoscope (fibroscope).

The focusing of light on the retina

Light enters the eye through the cornea, where most of the bending of the light occurs. The light then passes through the pupil, the size of which is controlled by the iris. The lens then helps to make any fine adjustments to the focusing of the light on the retina at the back of the eye. The light-sensitive cells located on the retina then send messages to the brain through the optic nerve.

The structure of the eye

1. The cornea is a tough curved transparent membrane over the front of the eye. It helps to focus light onto the retina.

2. The **eye lens** is a convex lens of variable thickness that focuses light to form a sharp image on the retina.

3. The **retina** is a layer of light-sensitive cells covering the inside of the back of the eye. When light falls on a retinal cell, an electrical impulse from the cell is transmitted to the brain via a nerve fibre. The brain interprets the pattern of impulses and recognises the image.

4. Transparent **fluids** fill the eyeball, keeping the internal structure under pressure and in place.

5. The **pupil** is the black circular aperture behind the cornea which light must pass through to reach the retina.

6. The **iris** controls the width of the pupil, which controls the entry of light into the eye. In bright light, the pupil automatically becomes narrow and therefore reduces the amount of light entering the eye. In dark conditions, the opposite happens.

The eye

Refraction

Refraction is the bending of light when it passes from one substance to another, e.g. from air to glass.

1. Light going through a **rectangular glass** block from air.

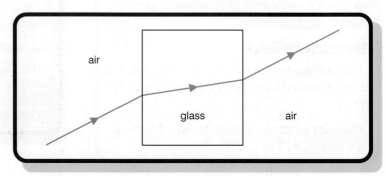

Light going from air to glass to air

54

2. Light passing through a **convex lens**, i.e. a lens which is thicker in the middle than at the edges.

Convex lens: light rays converge or come to a focus.

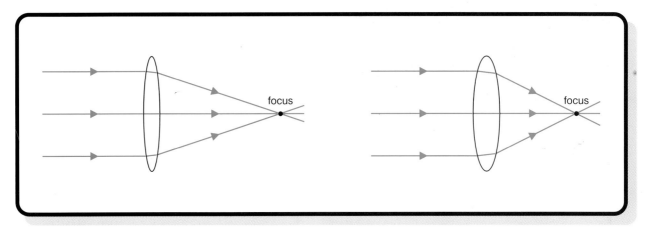

Thin convex lens – parallel light   Thick convex lens – parallel light

3. Light passing through a **concave lens**, i.e. a lens which is thicker at the edges than in the middle.

Concave lens: light rays diverge or spread out.

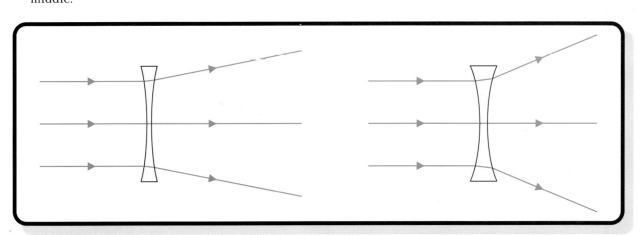

Thin concave lens – parallel light   Thick concave lens – parallel light

Note that the more convex or concave a lens is, the more the light bends.

The convex lens

★ The **focal point** of a convex lens is the point where light rays parallel to the lens axis are brought to a focus.

★ The **focal length** is the distance from the lens to the focal point.

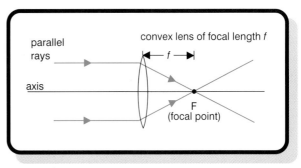

Focal length

Image formation on the retina

The image (or picture) formed by the retina of the eye is upside down and laterally inverted (rotated sideways).

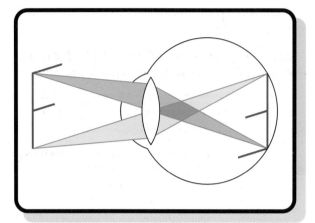

Image upside down and laterally inverted

This image formation can be explained using a ray diagram as shown below.

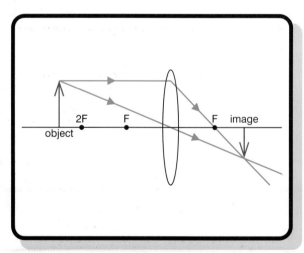

Ray diagram – formation of inverted image

To find the position of the image, a ray from the top of the object is drawn parallel to the axis of the lens. This ray bends through the focus. A second ray from the top of the object is drawn towards the centre of the lens. This ray does not bend at all.

Where the two rays from the top of the object cross over is the top of the image, which is therefore upside-down.

To find the focal length of a lens

The focal length of a lens can be found by focusing some distant object, such as a distant tree or building, on a wall of a room. The distance between the lens and the sharp image on the wall is the focal length of the lens.

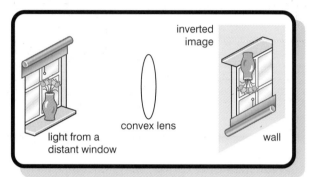

Focal length – pupil doing the experiment

Long and short sight

Both long and short sightedness are caused by the eye failing to bring the light rays to a focus on the retina.

A long sighted person can see only distant objects clearly. Images of nearby objects are formed behind the retina and appear blurred. This sight defect can be corrected by the use of spectacles with convex lenses.

A short-sighted person can only see nearby objects clearly. Images of a distant objects are formed in front of the retina and appear blurred. This sight defect can be corrected by the use of spectacles with concave lenses.

Question

1 Which type of lens, convex or concave, is used to correct **a)** short sight, **b)** long sight?

Fibre optics in medicine

An optical fibre is a thin fibre of transparent flexible material. A light ray that enters the fibre at one end emerges at the other end, even if the fibre is curved round. This happens because the light ray is totally internally reflected at the fibre

Answer 1 a) Concave b) Convex

surface wherever it hits the boundary. Each light ray in the fibre travels along a straight line through the fibre between successive reflections. Provided the bends in the fibre are not too tight, light rays in the fibre do not emerge from its sides.

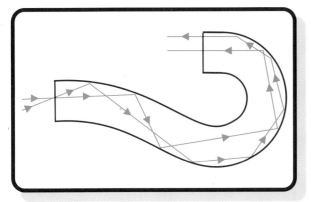

An optical fibre

In medicine, fibre optics are used in a device called an **endoscope** (see page 59) to examine the inside of a patient's body without the use of surgery, i.e. there is no need to cut open any part of the patient for examination.

The endoscope can be inserted down the throat of a patient, for example, and light is transmitted down the fibres. These fibres can transmit light without any heat coming from the lamp producing the light, i.e. the light is said to be 'cold'.

Refraction – angle of incidence, angle of refraction, normal

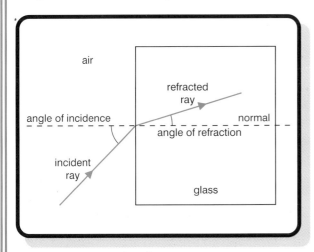

Refraction – incident ray, refracted ray

Normal: a line at right angles to the glass-air boundary where the beam of light hits the glass.

Angle of incidence: angle between incident ray and the normal.

Angle of refraction: angle between refracted ray and the normal.

Image formation of an object close to and far away from the eye

The eye lens is responsible for fine focusing and can change its shape to adjust for the distance between the object and the eye.

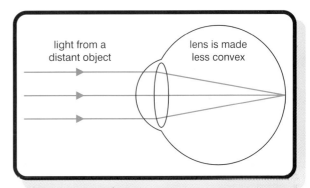

Distant object – thin eye lens

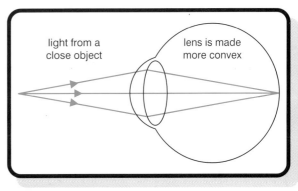

Close object – thick eye lens

Power of a lens

The power of a lens is a measure of how strong it is and is related to the focal length.

Power of a lens = 1/focal length or $P = 1/f$

Focal length = 1/power of lens or $f = 1/P$

Power is measured in dioptres (D) and focal length in metres (m). A convex lens has a positive power and a concave lens has a negative power, e.g. if a convex lens has a focal length of 50 cm then its power is found from

$P = 1/f$

$P = 1/0.5 = +2\,D$

or if a concave lens has a power of –2.5 D, then its focal length is found from

$f = 1/P$

$f = 1/{-2.5} = -0.4\,m$ or $-40\,cm$

Correction of long and short sight

Long sight: rays from a near object focus behind the retina. This is corrected by placing a convex lens in front of the eye to provide more bending power to the eye lens.

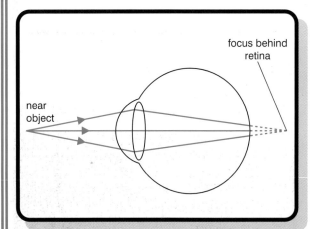

Long sight

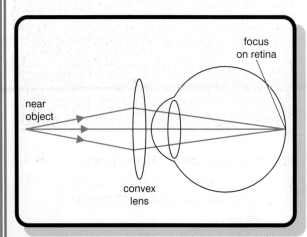

Correction to long sight

Question

2 a) Calculate the power of a concave lens of focal length 0.20 m
b) Calculate the focal length of a lens of power +4 D and state whether the lens is convex or concave.

Short sight: rays from a distant object focus in front of the retina. This can be corrected by placing a concave lens in front of the eye.

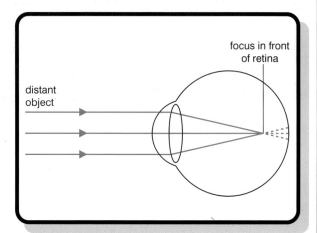

Short sight

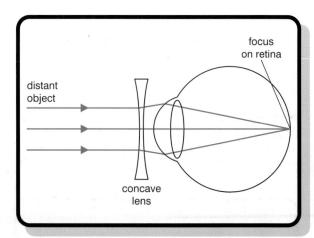

Correction to short sight

Answer

2 a) –5.0 D   b) 0.25 m, convex

Health physics

The endoscope

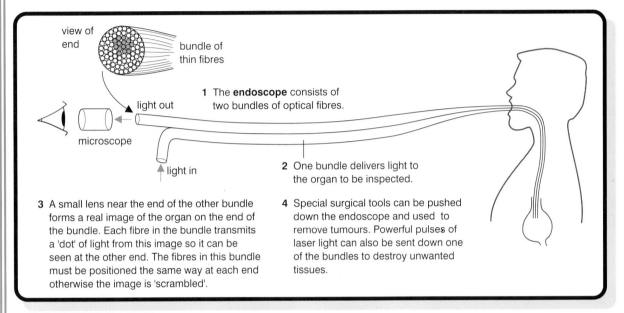

Doctors can put optical fibres inside a patient and examine internal organs such as the stomach. Such an instrument is called an endoscope or fibroscope and consists of two separate bundles of very thin glass fibres. One bundle shines the 'cold' light into the patient. The other bundle allows the doctor to see inside the patient either through an eyepiece or via a camera for viewing on a television screen.

Question

3 In an endoscope, why are there two bundles of optical fibres, not one?

Answer
3 One bundle is used to illuminate the internal cavity. The other bundle is used to transmit light out so the doctor can see inside the patient.

3.4 Using the spectrum

preview

At the end of this topic you will be able to:

- describe how the laser is used in one application of medicine
- describe one use of X-rays in medicine
- state that photographic film can be used to detect X-rays
- describe the use of ultraviolet and infrared in medicine
- state that excessive exposure to ultraviolet radiation may produce skin cancer
- describe one advantage of computerised tomography.

All electromagnetic waves

1. do not need to be carried by a medium
2. travel at the same speed of 300 000 km/s in a vacuum
3. can be diffracted, refracted and reflected (although X-rays and gamma rays need special techniques)

Revise Standard Grade Physics

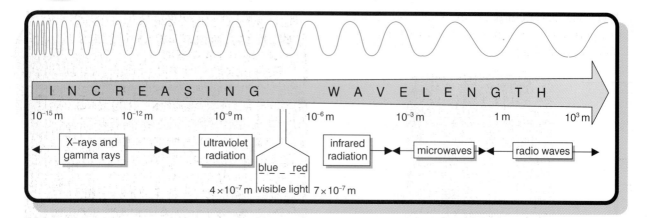

Lasers in medicine

A laser produces a very intense beam of light which can be used in various surgical techniques:

1. vaporising tumours and hence destroying them
2. sealing damaged blood vessels, especially in eye surgery
3. removing tattoos and birth marks.

The use of X-rays in medicine

Doctors use X-rays to see inside patients. Dentists use X-rays to examine teeth. X-rays pass through soft tissue like skin, fat and muscle but are absorbed by bone. Breaks in bones therefore show up as dark on a photographic plate. Organs such as the lungs, brain and gut can also be examined using X-rays.

Infrared and ultraviolet radiation

Infrared radiation: special cameras using heat-sensitive film can be used to take thermograms. These 'heat' pictures show hot and cold areas of patients, e.g. tumours are warmer than healthy tissue and poor blood circulation will result in affected parts of the body being colder than normal, giving off less infrared radiation.

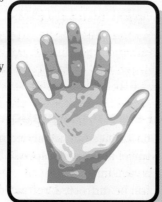

A thermogram

Ultraviolet radiation: long-stay patients in hospital are sometimes exposed to ultraviolet lamps to compensate for lack of sunlight. The ultraviolet radiation in sunlight produces vitamin D3 in our bodies, which is essential for life. Low doses of ultraviolet radiation are used to treat skin disorders. The ultraviolet radiation in sunlight gives us a suntan but overexposure is dangerous as it can damage the skin and increase the risk of skin cancer.

Questions

1. State one similarity and one difference between light and X-rays
2. a) Does infrared radiation have a shorter or a longer wavelength than ultraviolet radiation?
   b) Why is overexposure to ultraviolet radiation harmful?

Computer-assisted tomography (CAT)

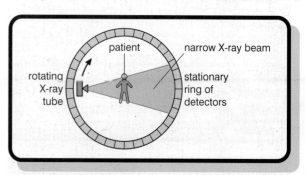

CAT scanner with patient

Answers
1 Both travel at the same speed through air; X-rays pass through tissue whereas light cannot. 2 a) Longer b) It causes skin cancer.

CAT uses X-rays but provides much more detail than ordinary X-ray photographs with the help of computer imaging.

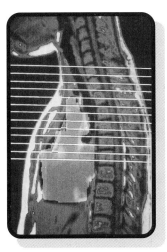

Photograph of a CAT scan

The X-ray tube and detector rotate around the body. Images of slices through the body are analysed by computer and displayed on a TV screen. One particular use of CAT scans is to detect brain tumours.

3.5 Nuclear radiation

preview

At the end of this topic you will be able to:

- state that radiation can kill living cells or change the nature of living cells
- describe one medical use of radiation based on the fact that radiation can destroy cells (instrument sterilisation, treatment of cancer)
- describe one medical use of radiation based on the fact that radiation is easy to detect
- state the range and absorption of alpha, beta and gamma radiation
- state that radiation energy may be absorbed in the medium through which it passes
- describe a simple model of the atom which includes protons, neutrons and electrons
- state that alpha radiation produces much greater ionisation density than beta or gamma radiation
- state one example of the effect of radiation on non-living things (e.g. ionisation, fogging of photographic film, scintillations)
- state that the activity of a radioactive source is measured in becquerels
- state that the activity of a radioactive source decreases with time
- describe the safety precautions necessary when dealing with radioactive substances
- state that dose equivalent is measured in sieverts
- explain the term ionisation
- describe how one of the effects of radiation is used in the detection of radiation (e.g. Geiger tube, film badges, scintillation counters)
- describe a method of measuring the half-life of a radioactive element
- state the meaning of the term 'half-life'
- carry out calculations to find the half-life of a radioactive source from appropriate data
- state that for living materials the biological effect of radiation depends on the absorbing tissue and the nature of the radiation, and that dose equivalent, measured in sieverts, takes account of the type and energy of radiation.

A scientific puzzle

Radioactivity was discovered in 1896 by **Henri Becquerel**. When he developed an unused photographic plate, he found an image of a key on it. He realised that this was caused by radiation from a packet containing uranium salts, which had been on top of a key with the photographic plate underneath. The puzzle of explaining the radiation was passed by Becquerel to a young research worker, **Marie Curie**. She painstakingly analysed the uranium salts and discovered the radiations were emitted from the uranium atoms, which formed other types of atoms in the process. She and her husband Pierre discovered and named two new radioactive elements, polonium and radium. It was shown that the emissions contained two types of radiation, alpha radiation which is positively charged and easily absorbed, and beta radiation which is negatively charged and much less easily absorbed. Later gamma radiation was discovered which is uncharged and much more penetrating.

Ernest Rutherford used alpha radiation to probe the atom. He knew that alpha radiation consisted of positively charged particles. He found that when a beam of alpha particles was directed at a thin metal foil, some of the particles bounced back off the foil. He deduced that

- the atom contains a tiny positively charged nucleus, where most of its mass is located
- the rest of the atom consists of empty space through which negatively charged electrons move as they orbit the nucleus.

Further investigations showed that the nucleus contains two types of particles, **protons** and **neutrons**.

Model of the atom

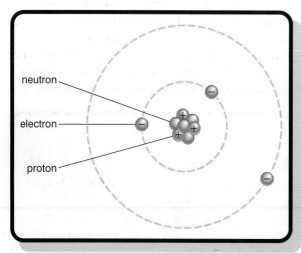

A lithium atom

An atom has a central core called the **nucleus** and inside the nucleus are tiny particles called **protons** and **neutrons**. The protons have a positive electric charge but the neutrons have no electrical charge.

Other tiny particles, much smaller than protons and neutrons, called **electrons**, move round the nucleus in orbits and these electrons have a negative charge. A neutral atom has an equal number of protons and electrons so the overall positive and negative charge cancels out.

Alpha particles are the most strongly ionising type of radiation because they knock electrons out of the atom much more effectively than either beta or gamma radiation. The correct order for causing ionisation is first alpha, then beta and finally gamma.

Range and absorption of alpha, beta and gamma radiation

Alpha particles are completely absorbed by a sheet of paper, by the outer layer of the skin or by a few centimetres of air.

Beta particles can pass through about 1 cm of body tissue, but are absorbed by a few millimetres of aluminium. They are thus more penetrating than alpha particles.

Gamma rays are the most penetrating type of radiation and can pass through the human body. They are absorbed by several centimetres of lead or about a metre of concrete.

When radiation passes through material, it loses energy by colliding with electrons in the material and knocking these electrons out of the atoms of the material. This is called ionisation. Energy is lost in the process of ionisation.

The properties of alpha, beta and gamma radiation

	alpha	beta	gamma
charge	+2	−1	0
absorption	thin paper	few mm of aluminium	several cm of lead
range in air	fixed, up to 10 cm	variable, up to 1 m	spreads without limit
ionising effect	strong	weak	very weak

Health physics

Nuclear radiation and living things

Nuclear radiation may damage or kill human body cells. The damage can be severe and cause immediate effects or it may be more subtle and have long-term effects. Cancers, cataracts and genetic damage are examples of long-term effects.

However, radiation therapy makes use of the short-term effects in medicine. The radiation is directed at cancerous cells in a tumour to kill or severely damage them. Radiation can also be used to sterilise surgical equipment since it kills any living bacteria present on these.

For some medical conditions, radioactive tracers can be injected into a patient. Since radioactive substances are easy to detect, their path around the body can be traced using special instruments. The radiation may also be concentrated in certain organs in the body and this helps doctors to diagnose or treat disease.

Effect of radiation on non-living things

Radiation can cause:

- ionisation
- 'fogging' of photographic film
- chemicals to emit flashes of light called scintillations.

Stable and unstable nuclei

A nucleus is **stable** if the strong nuclear force between its neutrons and protons is much greater than the electrostatic force of repulsion between the protons. Some nuclei are **unstable** because the electrostatic forces of repulsion are larger than the strong attractive forces.

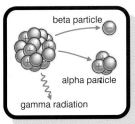

An unstable nucleus

★ A large nucleus with **too many protons** and **neutrons** is unstable. It becomes stable by emitting an **alpha particle**. This is a particle consisting of two protons and two neutrons.

★ A smaller nucleus with **too many neutrons** is unstable. It becomes stable by emitting a **beta particle**. This is an electron created in the nucleus and instantly emitted.

★ A nucleus may still possess **excess energy** after an alpha or beta particle has been emitted. It may then release the excess energy as **gamma radiation**. This is electromagnetic radiation of very short wavelength.

★ The daughter nucleus might itself be radioactive, and may emit a further alpha or beta particle.

★ An unstable nucleus is said to **disintegrate** when it emits an alpha particle or beta particle. When it emits gamma radiation, it is said to **de-excite**.

★ The **activity** of a radioactive source is the number of nuclei per second that disintegrate.

The Geiger counter

This consists of a Geiger tube connected to an electronic counter. Each particle from a radioactive source that enters the tube is registered on the electronic counter as one count. If the Geiger tube is pointed at a radioactive source, the activity of the source can be monitored by counting the number of particles entering the tube in a measured time interval and calculating the **count rate** (the number of counts per second), which is proportional to the activity.

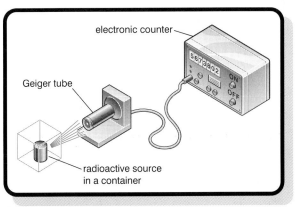

Using a Geiger counter

Background radioactivity

A Geiger counter will detect a low level of radioactivity even with no source present. This is called **background radioactivity** and is due to cosmic radiation and naturally occurring radioactive isotopes in rocks such as granite.

> **Question**

1. A Geiger counter records 1980 counts in 300 seconds when it is held at a fixed distance from a radioactive source. Without the source present, it records 120 counts in 300 seconds.
   a) Why does the Geiger tube count when no source is present?
   b) Calculate the count rate due to the source.
   c) Give two reasons why the count rate is less than the activity of the source.

Why radioactivity is harmful

Radiation from radioactive substances produces ions. Ionising radiation damages living cells in two ways:

1. by penetrating the cell membranes which causes the cell to die
2. by breaking the strands of DNA molecules in the cell nucleus, which may cause cell mutation.

* **Alpha radiation** is easily absorbed, highly ionising and therefore very harmful.
* **Beta radiation** is less easily absorbed and less ionising. However, it can penetrate deep into the body from outside so it too is very harmful.
* **Gamma radiation** easily penetrates soft tissue and is absorbed by bones, where its ionising effect can produce immense damage.

There is no lower limit below which ionising radiation is harmless. Therefore, extreme care is essential when radioactive substances are used and legal regulations for using radioactive substances must be observed. In a school laboratory, students under the age of 16 are not allowed to carry out experiments with radioactive materials.

> **Question**

2. a) Why is a storage box for radioactive substances made of lead?
   b) Why is it essential for handling tongs to have long handles?

Safety precautions

When handling radioactive substances:
1. always use forceps to lift a radioactive source
2. always point a radioactive source away from the body
3. never eat or drink where radioactive sources are being used
4. always wash your hands after handling radioactive sources
5. remember – people under 16 years of age must not handle radioactive sources.

Unit of activity of radiation

Each time an atom of a radioactive material gives out an alpha, beta or gamma particle the atom is said to have **disintegrated**. The number of disintegrations per second from a radioactive material is called its **activity** and this activity is measured in units called **becquerels** (Bq), named after the scientist who discovered radioactivity. Thus an activity of 350 Bq means that 350 atoms in the radioactive material disintegrate in one second. Other larger units are also used, e.g. kBq (1000 Bq) or MBq (1 000 000 Bq).

As atoms of a radioactive source disintegrate, there are fewer remaining and so there will be fewer disintegrations as time goes by. Thus the activity of a radioactive source decreases with time.

> **Answer**
>
> 1. a) background radioactivity
>   b) 6.2 counts per second
>   c) The radiation spreads out from the source in all directions so most of it misses the tube; absorption by the air between the tube and the source.

> **Answer**
>
> 2. a) Lead is the best absorber of radioactivity.
>   b) The tongs keep the source as far away from the user as possible.

Health physics

Dose equivalent

Radioactive materials emit penetrating radiation that can injure living tissue. The unit of radiation dose equivalent in human beings is the **sievert (Sv)**. This is a measure of the quantity of radiation absorbed by the body, corrected for the nature of the radiation, since different types are of different degrees of harmfulness.

The **millisievert (mSv)**, which is one thousandth of a sievert, is also commonly used. In the United Kingdom, each person is exposed to about 2.5 mSv per year from natural background radiation.

Ionisation

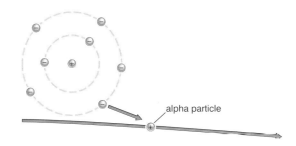

Ionisation

Ionisation is the process by which radiation knocks out an electron from an atom. The atom is initially neutral (same number of protons as electrons), but the removal of an electron means that the atom is no longer neutral – it is now called an ion and since there are now more positive charges than negative, the original atom is a positive ion. The single electron which was removed is now a negative ion, and so an ion pair is formed.

Radiation detectors

Geiger tube: radiation entering the very thin window of the tube ionises the gas inside and the ion pairs are attracted to either the positive wire or the negative tube.

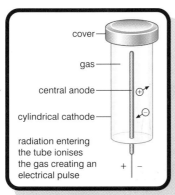

Geiger tube

This produces electrical pulses, which can be counted and displayed on a counter or heard as audible clicks on a speaker.

Photographic film: radiation blackens or fogs a piece of photographic film and the density of the fogging is a measure of the radiation received. Workers in the nuclear industry wear film badges which operate on the same principle.

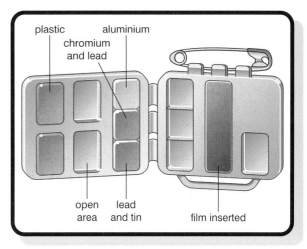

Film badge

Half-life

As the atoms of a radioactive material disintegrate or decay, there are fewer of them left and so as time goes by the activity of the material decreases. The **half-life** of a radioactive substance is the time taken for its activity to fall to half its original value.

Measurement of half-life

Half-life can be measured by placing a Geiger tube close to a source of radiation. The Geiger tube is connected to a counter and a stopwatch is used to record the time.

Before starting the experiment the background radiation must be measured. This can be found by removing the source from the vicinity of the experiment and measuring the number of counts in one minute from radiation in the room. It is best to repeat this several times as the number of counts per minute is random. This background reading must be subtracted from all subsequent readings when the source is in place.

With the source in place, readings of the count over a period of one minute is taken. Half an hour later, the number of counts per minute is taken again. After a further half hour the number of counts per minute is recorded and this process is continued for a few hours.

Finally, a graph is plotted of the corrected count rate (i.e. with background subtracted) against time. This is called a decay curve.

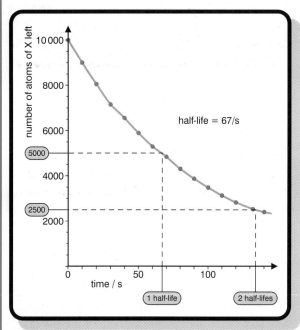

A decay curve

From the graph, the initial corrected count rate = 10 000 counts per minute

Half of this initial value = 5000 counts per minute

Time taken for the activity to fall to 5000 counts per minute = 67 s

So half-life of this source = 67 s

Note: the time taken for the activity to fall from 5000 counts per minute to 2500 counts per minute is also equal to the half-life of 67 s.

Calculations involving half-life

Worked example
What is the half-life of a radioactive substance if its activity falls from 1600 kBq to 100 kBq in 16 days?

Solution
Original activity = 1600 kBq

After one half-life, activity = 800 kBq

After two half-lives, activity = 400 kBq

After three half-lives, activity = 200 kBq

After four half-lives, activity = 100 kBq

Therefore from 1600 kBq to 100 kBq takes four half-lives, which is 16 days

Therefore half-life = 16/4 = 4 days

Worked example
A radioactive source has a half-life of 1200 years. How long will it take for the count rate to fall to $\frac{1}{8}$ th of its original value?

Solution
Let us assume that the original activity = 1

After one half-life, the activity = $\frac{1}{2}$ or $(\frac{1}{2})^1$

After two half-lives, the activity = $\frac{1}{4}$ or $(\frac{1}{2})^2$

After three half-lives, the activity = $\frac{1}{8}$ or $(\frac{1}{2})^3$

Therefore three half-lives = 3 × 1200 years

Therefore time taken = 3600 years

Factors affecting the biological effect of radiation

Radiation deposits energy into the living tissue it passes through.

The damage to tissue depends on:
1. the type of body tissue the radiation enters
2. the type of radiation concerned.

Using radioactivity

In each of the uses of radioactivity described below, think about the choice of the radioactive isotope in terms of

1. the half-life of the isotope
2. absorption of radioactive emissions
3. whether or not the daughter isotope is stable.

Medical uses

Tracers for diagnosis: the cause of certain illnesses can be pinpointed using **radioactive**

Health physics

tracers. For example, an underactive thyroid gland can be detected by giving the patient food containing the radioactive isotope iodine-131. This is a beta emitter with a half-life of 8 days. A correctly functioning thyroid gland will absorb iodine and store it. A Geiger tube pointed at the neck will therefore show an increased reading if the patient's thyroid gland is functioning correctly. The amount of radioactivity is too small to harm the gland, and the isotope decays after a few weeks.

Question

3 a) Why is a beta emitter chosen?
b) Why is an isotope with a half-life of a few minutes not chosen?

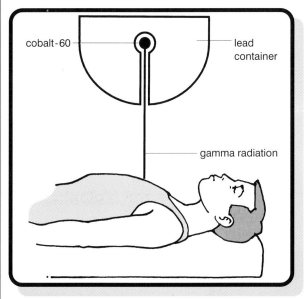

Treating cancer

Gamma therapy: gamma radiation from the radioactive isotope cobalt-60 is used to destroy cancerous tissues. The gamma radiation penetrates the body and passes into the diseased tissue. A lead collimator (filter) is used to direct the gamma radiation onto the cancer cells. The half-life of cobalt-60 is 5 years.

Question

4 a) Why is a gamma emitter chosen?
b) Why is an isotope with a half-life of a few years chosen?

Some other uses

Food radiation: gamma radiation is used to kill the bacteria responsible for food poisoning in certain foods. This makes the food safer to eat, and prolongs its shelf-life.

Answers

3 a) Beta radiation from the thyroid can be detected outside the body.
b) There would be no radioactivity left in the iodine by the time it reached the thyroid.
4 a) Gamma radiation easily penetrates the body.
b) The source only needs to be replaced every few years.

Equations review

$$P = 1/f$$

f here is the focal length of a lens (in metres, m) while P is the power of the lens (in dioptres, D). Positive values of P and f relate to convex lenses and negative values to concave lenses.

round-up

1 a) State the normal temperature in °C of the human body.
b) What feature of a clinical thermometer
  (i) makes it more sensitive than a laboratory thermometer?
  (ii) prevents its reading from falling after the thermometer has been removed from a patient? [3]

2 a) (i) Name a liquid that is used in a liquid-in-glass thermometer.
  (ii) Why is the stem of a clinical thermometer triangular in cross-section?
  (iii) What temperature range does a clinical thermometer cover?
b) What is a stethoscope used for in medicine? [5]

3 a) (i) Which one of the following frequencies is closest to the upper frequency limit of the normal human ear?

 A 20 Hz **B** 200 Hz **C** 2000 Hz **D** 20 000 Hz

 (ii) Which one of the following loudness levels is closest to the level you would be subjected to if you were in a very busy street with buses and lorries passing in both directions?

 A 1 dB **B** 10 dB **C** 100 dB **D** 1000 dB

b) State two uses of ultrasound in medicine. [4]

4 a) An ultrasonic cleaning tank operates at 4 MHz. Calculate the wavelength of ultrasonic waves in water. The speed of sound in water is 1500 m/s. [2]

b) When using an ultrasonic hospital scanner, a paste is applied to the body surface where the probe is used.

 (i) Explain why this is necessary [2]
 (ii) Give two reasons why each reflected pulse is weaker than the pulse emitted by the probe [2]

5 a) When a light ray passes from air into glass at a non-zero angle to the normal, does it bend away or towards the normal? [1]

b) A light ray passes from a pool of water into air, emerging at an angle of 30° to the vertical as in the following diagram. Is the angle between the light ray in the water and the vertical more than, less than or equal to 30°? [1]

c) In b) the angle between the light ray in the water and the vertical is increased until the refracted ray emerges along the water surface. What happens if the angle is made larger? [1]

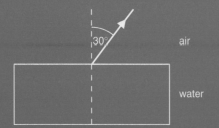

6 a) Copy and complete the diagram below to show the path of two light rays from a point object onto the retina of a corrected long-sighted eye.

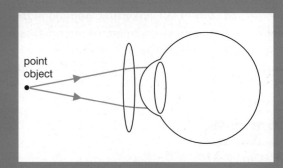

b) (i) What type of correcting lens is shown in the diagram above?
 (ii) The lens above has a focal length of 0.50 m. Calculate its power. [6]

7 a) A student sitting at the back of a class can read his book without the aid of spectacles but is unable to read writing on a board 5 m away. What is the name for this sight defect? [1]

b) With the aid of a diagram, explain the cause of this defect and describe how it can be corrected with a suitable lens. [3]

8 a) Explain with the aid of a diagram what is meant by total internal reflection of light. [2]

b) The diagram shows a light ray entering an optical fibre. Complete the diagram, showing the path of the light ray after it enters the optical fibre. [2]

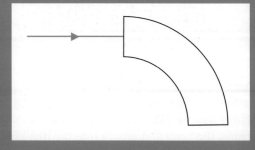

c) State two applications of optical fibres. [2]

9 Give one use in medicine of each of the following:
 a) X-rays
 b) ultrasonic radiation
 c) infrared radiation. [3]

10 State the three types of emissions from naturally occurring radioactive substances, and state which type of radioactive emission is most easily absorbed. [4]

11 In a test to identify the type of radioactivity produced by a radioactive source, the following results were obtained with different sheets of materials placed between the source and the Geiger tube.

Material	count rate/counts per second
none	450
tinfoil	235
1 mm aluminium	230
10 mm aluminium	228
10 mm lead	160

Use these results to decide what types of radiation are emitted by the source. Explain your answer [4]

12 a) Background radioactivity accounts for 87% of the exposure to ionising radiations of the average person in Britain. Explain what is meant by
  (i) background radioactivity [1]
  (ii) ionising radiation. [1]
 b) State two further sources of ionising radiation. [2]
 c) Explain how you would use a Geiger counter and a stopwatch to measure background radioactivity. [2]

13 The *half-life* of the *isotope* of carbon $^{14}_{6}C$ is 5500 years. Explain what is meant by the terms in italics. [2]

14 a) State the unit of (i) activity of a radioactivity source (ii) radiation dose equivalent of ionising radiation.
 b) In a radioactivity experiment, a Geiger tube was placed at a fixed distance from a radioactive source and the number of counts in one minute was measured as 540 counts. The measurement was repeated 30 minutes later using the same source at the same distance and found to be 400 counts. The source was removed and the background count was measured and found to be 25 counts per minute.
  (i) Calculate the corrected count for each measurement with the source present.
  (ii) Is the half-life of this source more than, less than or equal to 30 minutes? Give a reason for your answer. [6]

Total = 62 marks

Electronics

4.1 Overview

preview

At the end of this topic you will be able to:
- state that an electronic system consists of three parts: input, process and output
- distinguish between digital and analogue outputs
- identify analogue and digital signals from waveforms viewed on an oscilloscope.

Electronic systems

All electronic systems, no matter how simple or complicated, consist of three parts, namely **input**, **process** and **output**.

For example, for a stereo system:

input:	tape head or laser pickup
process:	amplifier
output:	loudspeakers

For a computer system:

input:	switches on keyboard
process:	decision making, calculations, etc.
output:	information on a visual display

Digital and analogue devices

Electrical signals are of two types, namely digital or analogue.

Digital signals have only two values: **on** (1) and **off** (0) (these are sometimes called **high** and **low**).

Analogue signals can have any value between a minimum and a maximum and can be changed gradually between minimum and maximum. Some examples of digital and analogue devices are listed below.

device	digital or analogue
switch	digital
dimmer switch	analogue
CD player	digital
loudspeaker	analogue
microphone	analogue
bulb	analogue or digital
liquid crystal display	digital
light-emitting diode	digital

Digital and analogue waveforms

Two electrical signals from an oscilloscope are shown below.

This one shows a digital signal: it is either high (1) or low (0)

This one shows an analogue signal: the signal varies continuously from a high positive value to a low negative value

Questions

1. Why is a light switch a digital device?
2. Why is a dimmer an analogue device?

Answers
1. A light switch is either on or off.
2. A light dimmer is used to smoothly change the brightness of a light bulb.

4.2 Output devices

preview

At the end of this topic you will be able to:
- give examples of output devices and the energy changes involved

- give examples of digital output devices and of analogue output devices
- draw and identify the symbol for an LED
- state that an LED will light only if connected one way round
- explain the need for a series resistor with an LED
- state that different numbers can be produced by lighting appropriate segments of a seven-segment display
- identify appropriate output devices for a given application
- describe by means of a diagram a circuit that will allow an LED to light
- calculate the value of the series resistor for an LED
- calculate the decimal equivalent of a binary number in the range 0000 to 1001.

Some output devices

Loudspeaker

A loudspeaker changes **electrical energy** into **kinetic energy** (of the loudspeaker cone) and to **sound energy**. It is used as an output device in electronic systems such as stereo systems, TVs and radios.

A loudspeaker is an analogue device.

Electric motor

An electric motor changes **electrical energy** into **kinetic energy**.

Electric motors are used as output devices in washing machines, vacuum cleaners, drills, etc.

The electric motor is an analogue device.

Relay

A relay is a switch operated by an electromagnet, i.e. when current passes through the electromagnet the relay switches on and when current stops flowing through the electromagnet the relay switches off. Relays in one circuit are used to activate devices in a second circuit that require larger currents or can be used to switch on a high voltage circuit using a low voltage circuit. A relay changes **electrical energy** into **kinetic energy**.

A relay is a digital device.

Solenoid

A solenoid again uses an electromagnet to produce movement. When the current is switched on, the plunger moves either in or out. A solenoid can be used in door chimes where a metal bolt moves to strike a plate giving out a chime or in central locking in a car. A solenoid changes **electrical energy** into **kinetic energy**. A solenoid is a digital device.

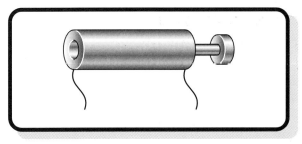

A practical solenoid

Light emitters: filament bulbs and light-emitting diodes (LEDs)

Filament bulbs and LEDs are useful output devices which convert electrical energy into light energy. The LED is a digital device whereas the bulb can be either digital or analogue.

In the circuit opposite, the bulb is connected directly across the voltage supply.

In the circuit on the following page, the LED is not connected directly across the voltage supply because the LED needs only a small current of about 10 mA to light, unlike the bulb which requires a much bigger current. To reduce the current, a resistor R is always used in series with an LED. Furthermore, the voltage across the LED should not exceed 2 V and the resistor overcomes this limitation by taking any excess of 2 V.

A bulb will light no matter which way round it is put into the circuit. This is not the case with an LED. It only works one way, as shown below.

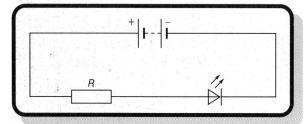

LED connected the right way

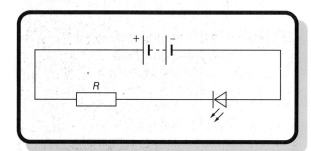

LED connected the wrong way

The seven-segment display

This is another light-emitting digital output device. It consists of seven LEDs arranged as a large figure of eight, as shown below.

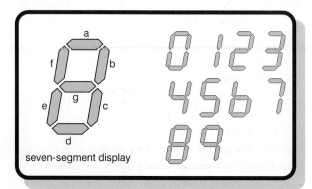

seven-segment display

By switching on different combinations of these LEDs, all the numbers from 0 to 9 can be formed.

Question

1 a) Sketch a circuit diagram to show a battery in series with a resistor and an LED in its forward direction.
b) A student connects an LED in series with a battery and a resistor but the LED does not light up. Suggest two possible reasons why the LED does not light up.

Selecting output devices

All output devices change an electrical signal from the process part of an electronic system into another form of energy. Some are digital, i.e. either on or off, and some are analogue, producing a range of output signals.

output action	possible output device needed
open or close curtains	motor
show a number	seven-segment display
ring a bell	solenoid
give a warning	LED
baby alarm	loudspeaker

A working LED circuit: series resistor value

The circuit diagram below shows an LED connected to a battery, which will allow the current to flow through the LED. A series resistor is necessary to prevent too much current flowing through the LED and causing damage. The calculation shown below allows the value of R to be determined.

Calculating the series resistance for an LED

An LED is used in a circuit with a 5 V supply.
The maximum LED voltage is 2 V and the current is 10 mA.
What value of series resistor is required?

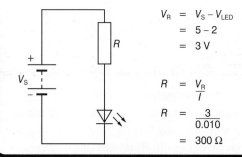

$$V_R = V_S - V_{LED}$$
$$= 5 - 2$$
$$= 3 \text{ V}$$

$$R = \frac{V_R}{I}$$
$$R = \frac{3}{0.010}$$
$$= 300 \, \Omega$$

Question

2 An LED lights when a voltage of 0.7 V is applied between its terminals. The LED is connected in series with a 1000 Ω resistor and a 3.0 V battery. Calculate
a) the voltage across the resistor
b) the current through the LED.

Converting binary numbers into decimal numbers

The binary system of counting is used in digital electronics. There are only two numbers in the binary system: 0 and 1. A typical binary number is 0110. Starting from the right-hand end, the first number indicates the number of 1s, the second column the number of 2s, the third column the number of 4s and the fourth column the number of 8s. So for 0110, there are zero 1s, one 2, one 4 and zero 8s. Thus the decimal equivalent is 2 + 4 = 6.

binary number	decimal equivalent
0000	0
0001	1
0010	2
0011	3
0100	4
0101	5
0110	6
0111	7
1000	8
1001	9

Answers

1 a) See the circuit diagram above.
b) The LED could be connected the wrong way round; the resistance of the resistor could be too large; the battery could be flat.
2 a) 2.3 V b) 0.0023 A

4.3 Input devices

preview

At the end of this topic you will be able to:

- describe the energy transformations involved in the following devices: microphone, thermocouple and solar cell
- state that the resistance of a thermistor changes with temperature and the resistance of an LDR decreases with increasing light intensity
- carry out calculations using $V = IR$ for thermistors and LDRs
- state that during charging the voltage across a capacitor increases with time
- identify from a list an appropriate input device for a given application
- carry out calculations involving voltages and resistances in a voltage divider
- state that the time to charge a capacitor depends on the values of the capacitance and the series resistance
- identify appropriate input devices for a given application.

Input devices

Microphone: a microphone changes sound energy into electrical energy. It can be used as an input device in public address systems or tape recorders. It is an analogue device.

microphone symbol

Thermocouple: a thermocouple consists of two wires of different metals twisted together to form a junction. When the junction is heated a small voltage appears across the two wires. It therefore changes heat energy into electrical energy and the hotter the junction, the greater the voltage produced. It is an analogue device.

thermocouple symbol

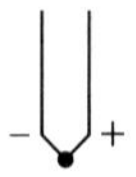

Revise Standard Grade Physics

Solar cell: a solar cell is made of a substance which produces a voltage when light falls on it, i.e. a device which changes light energy into electrical energy. Lots of solar cells are used to produce electrical energy in spacecraft. Photographers use solar cells in light meters to measure the light intensity. The more light there is, the greater the voltage produced. It is an analogue device.

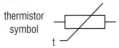
solar cell symbol

Thermistor: a thermistor is made of a material whose resistance changes with temperature. The hotter the thermistor, the smaller its resistance. A thermistor can be used as a kind of thermometer or as a temperature monitor and controller. It is an analogue device.

thermistor symbol

Light dependent resistor (LDR): an LDR is made of a material whose resistance decreases as the amount of light falling on it increases. It is an analogue device and is often used in automatic lighting controls and burglar alarm systems.

LDR symbol

Thermistor and LDR calculations

Keeping in mind that these devices are resistors (whose value can change depending on conditions), the equation $R = V/I$ can be used.

Worked example
At 20 °C, the voltage across a thermistor is 2 V and the current through it is 0.2 A. At 60 °C, the current is 0.4 A whilst the voltage remains the same. Calculate the resistance of the thermistor at each of these temperatures.

Solution
At 20 °C, $R = V/I = 2/0.2 = 10\,\Omega$

At 60 °C, $R = V/I = 2/0.4 = 5\,\Omega$

Note: for a thermistor, as $T \uparrow, R \downarrow$

Worked example
An LDR is connected across a 5 V supply. In daylight the current through the LDR is 0.1 A, whereas in the dark the current drops to 0.01 A. Find the resistance of the LDR in light and dark conditions.

Solution
Light $\qquad R = V/I = 5/0.1 = 50\,\Omega$

Dark $\qquad R = V/I = 5/0.01 = 500\,\Omega$

Note: for an LDR, as the light level $\uparrow, R \downarrow$

Question

1 a) A thermistor at a certain temperature was connected in series with a 50 Ω resistor, an ammeter and a 4.5 V battery. The ammeter reading was 0.020 A. Calculate **(i)** the voltage across the resistor **(ii)** the resistance of the thermistor.
b) Explain why the ammeter reading increases when the thermistor is warmed up.

Capacitors

Capacitors are electrical components which **store** electric charge. As soon as a capacitor begins to store some charge, a voltage appears across the terminals of the capacitor. The more charge held, the greater this voltage.

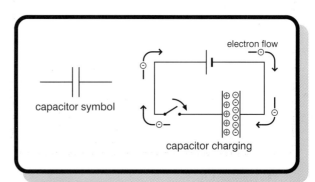
The capacitor

Answer 1 a) (i) 1.0 V (ii) 175 Ω b) The resistance of the thermistor decreases when it is warmed up. The total circuit resistance therefore decreases, allowing the current to increase.

Charging a capacitor

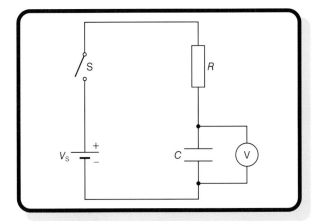

Charging a capacitor

When switch S is closed, charge flows into the capacitor and as the amount of charge builds up, the voltage shown by the voltmeter increases.

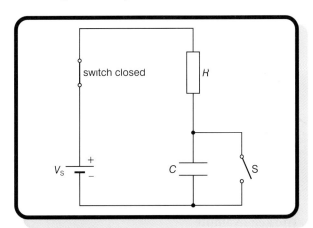

Discharging a capacitor

By connecting one side of the capacitor directly to the other, the capacitor discharges very quickly.

Question

2 A capacitor was connected in series with a battery, an LED and a switch. When the switch was closed, the LED lit up briefly. Explain why **a)** the LED lit up when the switch was closed **b)** the LED did not stay lit.

Selecting input devices

All input devices transform some form of energy into electrical energy. The electrical signal can then be passed to the process part of the electronic system. In deciding which input device to use in a system, the first question to be answered is 'what kind of energy does the system have to detect?'

If the answer is light, then the input device could be an LDR or solar cell. If it is heat, the input device could be a thermocouple or thermistor. If it is sound, then the input device could be a microphone and so on.

The potentiometer and the voltage divider

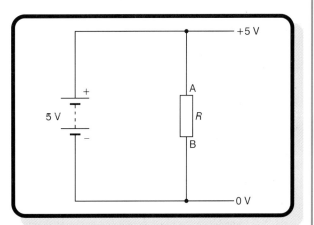

Voltage across a resistor

In this simple circuit, resistor AB is connected across a 5 V supply. The voltage across AB is 5 V. Assuming the negative side of the supply is 0 V, then the other side of the supply is +5 V. Hence the voltage at A is +5 V and the voltage at B is 0 V.

Answer

2 a) Closing the switch allowed the capacitor to become charged. The flow of charge passing through the capacitor also passed through the LED so the LED lit up.
b) The flow of charge to the capacitor stopped when the capacitor voltage became equal to the battery voltage. The LED only lit up when charge was flowing through it.

Revise Standard Grade Physics

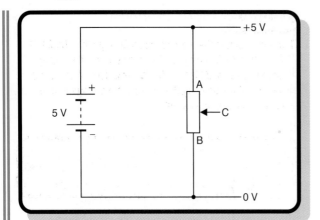

The potentiometer

If the resistor above is replaced by a potentiometer, then the voltage at A is still +5 V and the voltage at B is still 0 V. However, the voltage at point C will vary between 0 V and +5 V depending on where point C is. When point C is very close to A the voltage at C will be almost +5 V and when point C is very close to B, the voltage at point C will be very nearly 0 V.

The voltage across part CB of the potentiometer will change from +5 to 0 V, while the voltage across part AC of the potentiometer will vary from 0 to +5 V.

The voltage divider

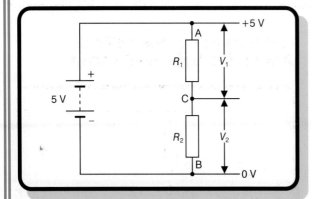

The voltage divider

The potentiometer above can be replaced by two single resistors R_1 and R_2 of fixed values. The voltage at A is +5 V and the voltage at B is 0 V. However, the voltage at C will depend on the value of the two resistors.

The current I is the same through each resistor since they are in series and since from Ohm's law, $I = V/R$

$I = V_1/R_1 = V_2/R_2$

therefore $V_1/V_2 = R_1/R_2$

In other words the ratio of the voltages across the resistors is the same as the ratio of the resistors.

Worked example

Two $1\,k\Omega$ resistors in series are connected to a supply voltage of 5 V. Calculate the voltage across each resistor.

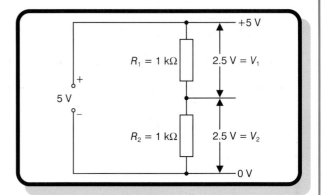

Solution
Since the resistors are equal, the ratio of R_1/R_2 is $1\,k/1\,k$ or simply 1. Therefore the ratio of the voltages $V_1/V_2 = 1$ and so $V_1 = V_2$. The supply voltage therefore splits equally across both resistors and so the voltage across each is 2.5 V.

Worked example
A $1\,k\Omega$ resistor and a $3\,k\Omega$ resistor in series are connected to a supply voltage of 8 V. Calculate the voltage across each resistor.

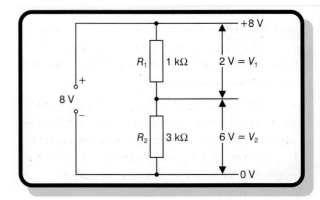

Solution

The ratio of the resistors is 1:3 and so the ratio of the voltages across them is 1:3, i.e. $V_1/V_2 = 1/3$, so $V_2 = 3V_1$. The 8 V must therefore be split across the two resistors so that the voltage across R_2 is three times the voltage across R_1. The only way this can be done is if V_2 is 6 V and V_1 is 2 V.

An alternative to this method is as follows:

$V_{total} = 8\,V$

$R_{total} = 4\,000\,\Omega$

so $I_{total} = V/R = 8/4\,000 = 0.002\,A$

$V_1 = I_{total} R_1 = 0.002 \times 1000 = 2\,V$

$V_2 = I_{total} R_2 = 0.002 \times 3000 = 6\,V$

Question

3 In the circuit in the worked example opposite, the 3 kΩ resistor is replaced by a 2 kΩ resistor and the voltage supply is replaced by a 9.0 V battery. The other resistor in the circuit is unchanged. Calculate the voltage across each resistor in the circuit.

Answer

3 The voltage across the 2 kΩ resistor = 6.0 V, the voltage across the 1 kΩ resistor = 3.0 V.

Practical use of the voltage divider: the temperature sensor

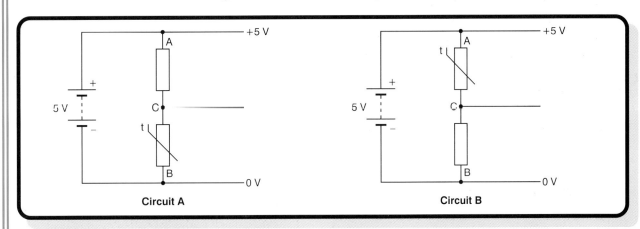

Circuit A   Circuit B

The resistance of a thermistor **increases** as the temperature **decreases**. In circuit **A**, as the temperature decreases, the resistance of the thermistor increases and therefore the voltage V_{BC} across it increases. V_{BC} is the output voltage of a voltage divider, i.e. this is the voltage which is passed on to the next process stage. If the output voltage V_{BC} rises, then this can be used to trigger off an alarm system, for instance giving warning of a decrease in temperature.

In circuit **B** above, as the temperature rises, the resistance of the thermistor decreases and so the voltage V_{AC} across it decreases. This means that the output voltage V_{BC} increases and so again can trigger an alarm system, warning of a high temperature.

Practical use of the voltage divider: the light level sensor

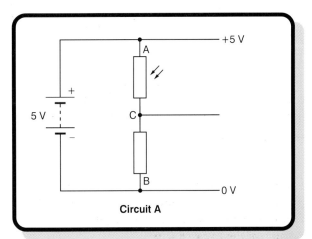

Circuit A

Revise Standard Grade Physics

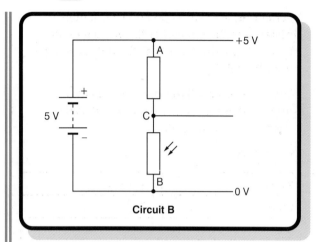

Circuit B

The resistance of an LDR **decreases** as the light level **increases**. In circuit **A** (page 77), as the light intensity increases, the resistance of the LDR decreases and so the voltage V_{AC} across it decreases. This means that the output voltage V_{BC} increases and triggers a warning of high light level.

In circuit **B**, as the light level decreases the resistance of the LDR increases and so the voltage V_{BC} across it will increase. Thus the output voltage increases, giving warning of a low light level.

If the light level in circuit **B** now increases, the resistance of the LDR will decrease, the output voltage V_{BC} will decrease and the warning system will be de-activated.

Thermistor and LDR resistance values

An easy way of remembering the resistance values of a thermistor and an LDR is to think about your reluctance or resistance to getting out of bed on a cold, dark morning compared to a hot bright morning.

Thermistor

temperature	your resistance	thermistor resistance	thermistor voltage
cold	high	high	high
hot	low	low	low

LDR

light level	your resistance	LDR resistance	LDR voltage
dark	high	high	high
light	low	low	low

Question

4 In the light level sensor circuit **A** (page 77) the light level is reduced by covering the LDR. Explain how this change will affect the voltage **a)** between A and C **b)** between B and C.

Altering the charging time in a capacitor circuit

The larger the value of R, the longer it takes to charge the capacitor C. The larger the value of C, the longer it takes to charge the capacitor C. The greater the value of the product $R \times C$, the longer the time required to charge the capacitor. This type of circuit is used in timing.

Selecting input devices

All input devices transform some form of energy into electrical energy. The electrical signal can then be passed to the process part of the electronic system. In deciding which input device to use in a system, the first question to be answered is 'what kind of energy does the system have to detect?' Below are some examples of selected input devices for given applications.

application	selected input device
light meter	solar cell
public address system	microphone
streetlight switch	LDR
time delay circuit	capacitor (+ resistor)
speed control	potentiometer
electronic thermometer	thermistor

Answer

4 a) The voltage between A and C will increase because the resistance of the LDR will decrease as a result of covering it. **b)** The voltage between B and C will decrease because it is equal to the supply voltage less the voltage between A and C.

The switch as a voltage input device

A switch can be used to supply an output signal which is either high or low when the switch is closed.

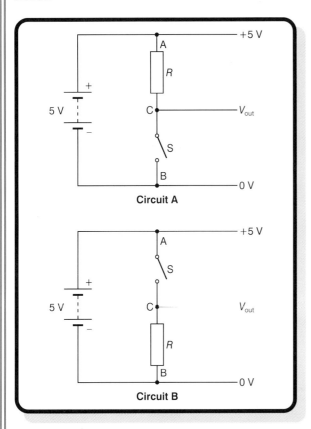

Circuit A

Circuit B

Treat an open switch as having a huge resistance (since no current flows through an open switch) and a closed switch as having zero resistance (because it then behaves as a resistanceless wire).

In circuit **A**, when S is open it is similar to a massive resistor in the bottom half of the potential divider. This will take all 5 V and so the input voltage V_{BC} is 5 V. When S is closed there is no resistance in the bottom half of the potential divider so the voltage V_{BC} is zero.

In summary:

switch	$V_{BC} = V_{out}$
closed	0 (low)
open	1 (high)

In circuit **B**, when S is open there is a huge resistance in the top half of the voltage divider which will take all 5 V and so the output voltage across BC will be zero.

When S is closed, all the resistance is in the bottom half of the voltage divider and so the output voltage will be 5 V.

In summary:

switch	$V_{BC} = V_{out}$
closed	1 (high)
open	0 (low)

Question

5 In the switch circuit **A** above, explain why the output voltage is 5 V when the switch is open.

Answer

5 No current passes through this circuit when the switch is open therefore the voltage across resistor R is zero. The voltage across the switch is equal to the supply voltage less the voltage across resistor R so the voltage between the output and the zero volt terminal is 5 V when the switch is open.

4.4 Digital processes

preview

At the end of this topic you will be able to:
- state that a transistor can be used as a switch
- state that a transistor may be conducting or non–conducting, i.e. on or off
- draw and identify the circuit symbol for an NPN transistor
- identify from a circuit diagram the purpose of a simple transistor switching circuit
- draw and identify the symbols for two input AND and OR gates and a NOT gate
- state that logic gates may have one or more inputs and that a truth table shows the output for all possible input combinations
- state that high voltage = logic 1 and low voltage = logic 0

Revise Standard Grade Physics

- draw truth tables for two input AND and OR gates and a NOT gate
- explain how to use combinations of digital logic gates for control in simple situations
- state that a digital circuit can produce a series of clock pulses
- give an example of a device containing a counter circuit
- state that there are circuits which can count digital pulses
- state that the output of the counter circuit is binary
- state that the output of the binary counter can be converted to decimal
- explain the operation of a simple transistor switching circuit
- identify the following gates from truth tables: two-input AND, two-input OR and NOT (inverter)
- complete a truth table for a simple combinational logic circuit
- explain how a simple oscillator built from a resistor, capacitor and inverter operates
- describe how to change the frequency of a clock pulse generator.

through the base to the collector but only when the voltage at the base is $\geq +0.7$ V. When this happens the transistor is said to be on. When the voltage at the base is below $+0.7$ V the transistor does not conduct and the transistor is said to be off. Thus the transistor is behaving like a switch.

symbol for NPN transistor

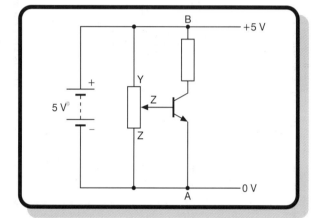

Transistor as a switch

In the circuit above, as the sliding contact Z of the potentiometer is moved from X to Y, the voltage at Z and therefore at the base of the transistor increases. When it reaches $+0.7$ V the transistor switches on and therefore current passes from A through the base and up through the collector to B, lighting the LED at the same time.

Transistor-switching circuits

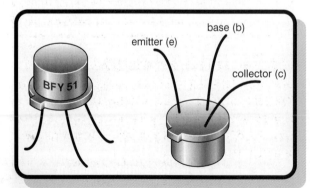

Transistors

The NPN transistor has three terminals and these are called the **emitter**, the **base** and the **collector**. Note that the arrow merely signifies the emitter terminal and does not indicate the direction of current flow. Electrons flow from the emitter

Temperature-controlled switch

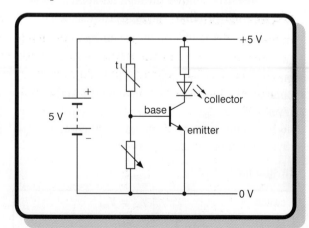

In the circuit at the bottom of page 80, when the temperature rises, the resistance of the thermistor decreases so the voltage across it decreases. This means that the voltage across the lower resistor increases. When the voltage reaches + 0.7 V the transistor conducts and the LED lights up.

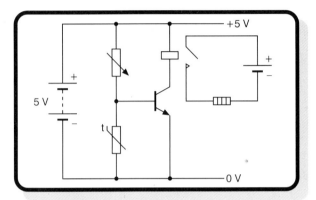

In the circuit above, when the temperature decreases, the resistance of the thermistor increases and so the voltage across it increases. When it passes + 0.7 V the transistor again conducts and switches on the relay, which operates a heater. When the temperature rises again, the resistance of the thermistor decreases so the voltage across it decreases and when it falls below + 0.7 V the transistor switches off, the relay switches off and finally the heater switches off.

Note: in both circuits, the temperature at which the transistor triggers is controlled by the setting of the variable resistor.

Light-controlled switch

In the circuit below, when the LDR is in the light, its resistance decreases and so the voltage across it will decrease. The voltage across the bottom resistor will then increase and the transistor will switch on.

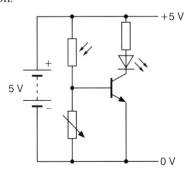

In the circuit below, when the LDR is in dark surroundings, its resistance will increase so the voltage across it will increase and the transistor will switch on the relay and operate the light.

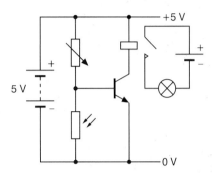

Once again, in both circuits, the value of the variable resistor will determine at what light level the transistor operates.

Time-delay switch

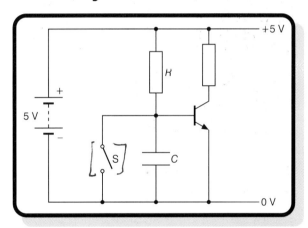

In this circuit, when the push switch is pressed the capacitor discharges and so the voltage across it is zero. When the push switch is released, the capacitor begins to charge up, so the voltage across it begins to rise from zero. When it reaches + 0.7 V the transistor switches on.

The time delay before the light switches on depends on the values of C and R. The bigger the values of C and R, the longer the time delay.

This type of circuit could be used to control lights at a pedestrian crossing.

Revise Standard Grade Physics

> **Question**
>
> **1** In the time-delay circuit shown on page 81
> a) why is the capacitor discharged as a result of pressing the push switch?
> b) why does the LED go off when the push switch is pressed?
> c) why does the LED not light up immediately after the push switch is released?

Logic gates and states

A **logic gate** is an example of a **digital circuit**. This is a circuit in which the voltage is either high (1) or low (0). No other voltage level occurs. There can only be two states (1 or 0) for the voltage at any point.

The output state of a logic gate is determined by the input states. A **truth table** is a table showing the output state for all possible input states.

The **NOT gate** is sometimes called an **inverter**. Its output state is NOT the same as its input state.

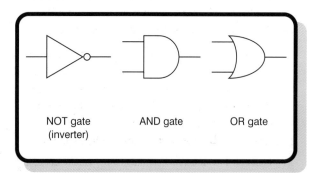

NOT gate (inverter)   AND gate   OR gate

The AND gate

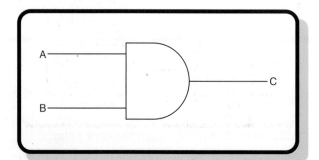

In an AND gate, the output is high only if input A **and** input B are both high. The following table is the truth table for an AND gate.

A	B	C
0	0	0
0	1	0
1	0	0
1	1	1

The OR gate

In an OR gate the output is high if either input A is high or input B is high (or both). The truth table for an OR gate is shown below.

A	B	C
0	0	0
0	1	1
1	0	1
1	1	1

Combinations of logic gates

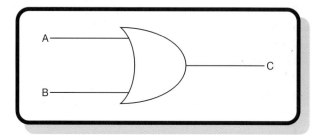

Gates can be used to switch things on or off. This circuit will switch on an LED when it gets dark. With less light, the resistance of the LDR increases so the voltage across it increases. This means that

> **Answer**
>
> **1 a)** The switch joins the capacitor terminals together when it is pressed.
> **b)** The base emitter voltage is zero so the transistor is off.
> **c)** The LED does not light up until the base emitter voltage reaches 0.7 V, i.e. until the capacitor charges to 0.7 V.

the voltage across the bottom resistor decreases. The input to the NOT gate will therefore switch from high to low and the output of the NOT gate will switch from low to high, thus switching on the LED.

By adding an AND gate to the original circuit, the output will change from 0 to 1 when it gets dark, but only when control switch S is also closed.

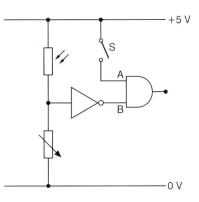

Question

2 The AND gate in the circuit above is replaced by an OR gate
  a) Sketch this new circuit.
  b) What difference will this change make to the operation of the circuit?

Clock pulse generator (oscillator)

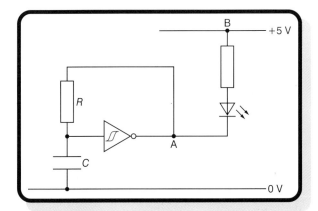

The clock pulse generator uses a special type of NOT gate with a special type of input called a Schmitt input.

The LED will light *only* when the logic level at A is 0 since a voltage of 5 V will appear across the LED. If the logic level at A is high (5 V) there will be no voltage across the LED and it will not light.

An oscillator produces digital clock pulses, i.e. a series of on (1) and off (0) signals. The LED in the above circuit will therefore flash on and off at a frequency determined by the values of R and C.

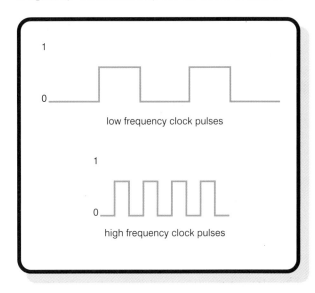

low frequency clock pulses

high frequency clock pulses

Any electronic system requiring a timing device will contain a clock pulse generator, e.g. traffic lights, microwaves, washing machines, etc.

Clock pulses can be counted by a binary counter circuit containing four LEDs which light up in a special sequence to show binary numbers in turn from 0000 to 1001, i.e. decimal 0 to 9. The output of the counter is said to be binary since it indicates numbers based on the binary system.

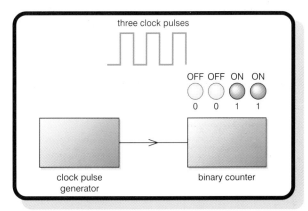

Answer

2 a) See the circuit at the top of the left-hand column, with an OR gate instead of an AND gate.
b) The output will change from 0 to 1 when it gets dark or when the switch is pressed.

Counting pulses in decimal

If the output from a binary counter is fed into a seven-segment display via a decoder, the binary numbers are converted to decimal by the decoder and then displayed on the seven-segment display.

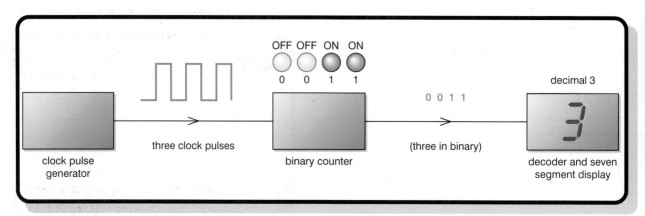

Transistor-switching circuits

To switch on a transistor, a voltage $\geq +0.7\,\text{V}$ must be applied between the emitter and the base. This is sometimes referred to as the base-emitter voltage or V_{eb}.

if $V_{eb} \geq +0.7\,\text{V}$ the transistor is switched on

if $V_{eb} < +0.7\,\text{V}$ the transistor is switched off

V_{eb} can be controlled by having components such as LDRs, thermistors, variable resistors and capacitors between the emitter and the base (or between the collector and the base).

Identifying gates from truth tables

input	output
0	1
1	0

NOT gate

A	B	C
0	0	0
0	1	0
1	0	0
1	1	1

AND gate

A	B	C
0	0	0
0	1	1
1	0	1
1	1	1

OR gate

An input of zero volts is called '0 V' or 'low' or 'logic 0'. An input of + 5 V is called 'high' or 'logic 1'.

Combinational logic circuit

Gates can be combined to give different truth tables, e.g. an OR gate can be combined with a NOT gate.

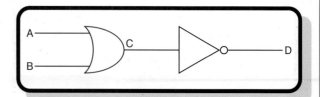

The truth table is as follows:

A	B	C	D
0	0	0	1
0	1	1	0
1	0	1	0
1	1	1	0

The clock pulse generator: how it works

Initially the capacitor is uncharged so the voltage across it is zero therefore the input to the NOT gate is 0, and the output from the NOT gate is therefore 1. The capacitor then begins to charge up through the resistor R and the voltage across it increases. At a certain value (V_1) the input to the gate is 1 and so the output is 0.

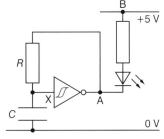

Clock pulse generator

The capacitor then begins to discharge through resistor R so the voltage across it drops until it falls to a lower value than V_1, which is necessary to return the output to 1.

The whole process then begins all over again. The frequency of the pulses is determined by the values of C and R. High values of C and R give low frequencies and low values of C and R give high frequencies.

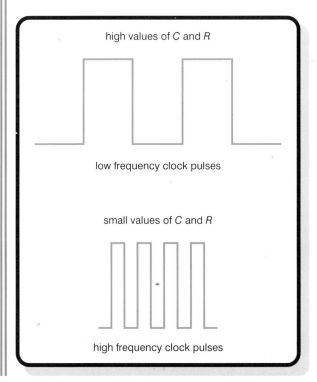

high values of C and R

low frequency clock pulses

small values of C and R

high frequency clock pulses

Question

3 A pulse generator produces pulses at a certain rate. How will the pulse frequency change as a result of **a)** reducing the resistance R **b)** increasing the capacitance C?

Answer

3 a) The frequency will increase.
b) The frequency will decrease.

4.5 Analogue processes

preview

At the end of this topic you will be able to:
- identify from a list devices in which amplifiers play an important part
- state the function of the amplifier in devices such as radios, intercoms and music centre
- state that the output signal of an audio amplifier has the same frequency as, but larger amplitude than, the input signal
- carry out calculations involving input voltage, output voltage and voltage gain of an amplifier
- describe how to measure the voltage gain of an amplifier
- state that power may be calculated from V^2/R where V is the voltage and R the resistance (impedance) of the circuit
- state that the power gain of an amplifier is the ratio of power output to power input
- carry out calculations involving the power gain of an amplifier.

Amplifiers

Amplifiers are found in many electronic systems, especially those we listen to, e.g. hi-fi systems, tape recorders, baby alarms and radios. The function of an amplifier is to convert a weak signal into a stronger signal, which can then be fed to a loudspeaker. The energy required to amplify the signal comes from an amplifier power supply.

Revise Standard Grade Physics

In a good quality amplifier the shape of the output signal is identical to the shape of the input signal but the amplitude has been enlarged. There is **no** alteration in the frequency.

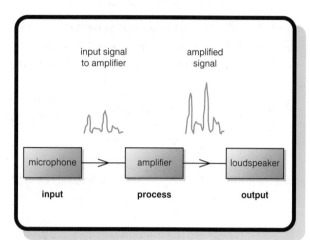

Voltage gain

The voltage gain of an amplifier can be calculated from:

$$\text{voltage gain} = \frac{\text{voltage output}}{\text{voltage input}}$$

*Note: voltage gain is simply a number with **no** units.*

Question

1 An amplifier used in a public address system at a railway station changes an input voltage of 0.05 V to an output voltage of 30 V. What is the voltage gain?

Measuring voltage gain

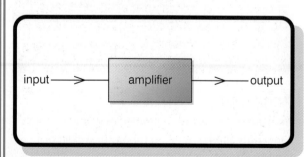

By connecting an oscilloscope first across the input of the amplifier and then across the output of the amplifier, both the input and output voltages can be measured and the voltage gain calculated.

Worked example

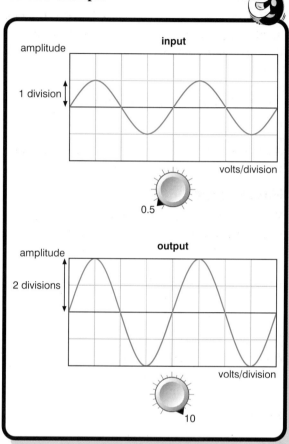

Voltage gain calculation:

The amplitude of the first trace is 1 division

Number of volts per division is 0.5

Thus the input voltage is $1 \times 0.5 = 0.5\,\text{V}$

Amplitude of output voltage is 2 divisions

Number of volts per division is 10

Thus the output voltage is $2 \times 10 = 20\,\text{V}$

$$\text{voltage gain} = \frac{\text{voltage output}}{\text{voltage input}} = \frac{20}{0.5} = 40$$

Note: this could also be done using a voltmeter in place of the oscilloscope.

Answer

1 Voltage gain = $\frac{\text{voltage output}}{\text{voltage input}} = \frac{30}{0.05} = 600$

Electronics

Power

Power (W) can be measured by using $P = IV$, but another equation can be derived by combining this with Ohm's law, i.e. $V = IR$.

$P = IV$
$= \frac{V}{R} \times V$ (since $I = \frac{V}{R}$ from Ohm's law)
$P = \frac{V^2}{R}$

*Note: R is sometimes referred to as **impedance** but this is still measured in ohms (Ω).*

Power gain

When dealing with hi-fi equipment, it is often more useful to consider the power gain of the amplifier rather than the voltage gain

$$\text{power gain} = \frac{\text{power output}}{\text{power input}}$$

*Note: power gain is simply a number with **no** units.*

Worked example

An amplifier has an input power of 200 mW and an output power of 50 W. Calculate the power gain of the amplifier.

Solution

$\text{power gain} = \frac{\text{power output}}{\text{power input}} = \frac{50}{0.2} = 250$

Worked example

A loudspeaker has a resistance of 6 Ω and a voltage of 12 V across it.

a) Calculate its power output.
b) If the input power of the amplifier is 10 mW, find the power gain of the amplifier.

Solution

a) power output = $\frac{V^2}{R} = \frac{(12)^2}{6} = \frac{144}{6} = 24$ W
b) Power gain = $\frac{\text{power output}}{\text{power input}} = \frac{24}{0.01} = 2400$

Question

2 A loudspeaker of resistance 5 Ω is connected to the output of an amplifier which has a power gain of 200. A signal of power 25 mW is applied to the input of the amplifier. Calculate **a)** the power output of the amplifier **b)** the current passing through the loudspeaker.

Equations review

1 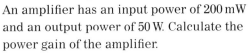 Voltage gain = voltage output / voltage input — Voltage gain has **no** units

2 $P = \frac{V^2}{R}$ — P is the power in watts (W), V is the voltage in volts (V) and R is the resistance or impedance in ohms (Ω).

3 Power gain = power output / power input — Power gain has **no** units

round-up

1 Which of the following devices are digital and which are analogue?
a) CD player b) light-emitting diode
c) loudspeaker d) solenoid [4]

2 A thermistor is connected in series with a resistor, a milliammeter and a 1.5 V cell.
a) Draw the circuit diagram for this arrangement.
b) When the thermistor is at 20 °C, the milliammeter reading is constant. State and explain how the milliammeter reading would change if the temperature of the thermistor fell. [4]

3 A capacitor, a switch, a light-emitting diode (LED) and a 1.5 V cell are connected in series with each other. When the switch is closed, the LED lights up briefly.
a) Sketch the circuit diagram for this arrangement.
b) Explain why the LED lights up briefly when the switch is closed. [3]

4 The circuit diagram shows a potential divider used to supply a variable voltage from a 9.0 V battery. The potential divider consists of a variable resistor R and a 1000 Ω resistor.
a) If the resistance of the variable resistor R is increased, what change occurs in

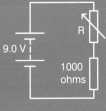

Answer
2 a) 5W
b) 1A

(i) the current from the battery
(ii) the voltage across the 1000 Ω resistor
(iii) the voltage across the variable resistor?
b) Calculate the voltage across the variable resistor when it is adjusted to 500 Ω. [5]

5 The circuit diagram shows a transistor being used to make a light-operated electric bell.

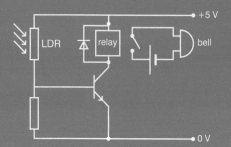

When the light-dependent resistor (LDR) is in darkness, the bell is off. When the LDR is illuminated, the bell rings. Explain why illuminating the LDR makes the bell ring. [4]

6 The diagram shows a potential divider circuit consisting of a light-dependent resistor (LDR) in series with a 1000 Ω resistor R and a 5.0 V voltage supply unit. A voltmeter is connected across resistor R.

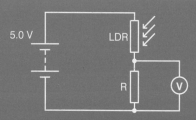

a) When the LDR is in darkness, the voltmeter reads 1.0 V.
(i) What is the voltage across the LDR when the voltmeter reads 1.0 V?
(ii) Calculate the resistance of the LDR when the voltmeter reads 1.0 V.
b) State and explain what change occurs in the voltmeter reading when the LDR is exposed to daylight. [6]

7 The diagram below shows a temperature-controlled heater. When the temperature of the thermistor falls, the heater is switched on.
a) Explain in terms of the current in each part of the circuit why the heater is switched on when the temperature of the thermistor falls. [3]

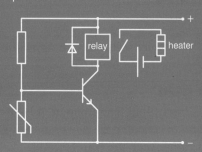

b) (i) What is the name of the component in parallel with the relay?
(ii) In which direction can electrons pass through this component? [2]

8 a) Identify the logic gate shown.
b) Write the truth table for this logic gate. [5]

9 The diagram shows an LED and a resistor used as a logic indicator connected to the output of a NOT gate.
a) State and explain what the logic state of the NOT input should be for the LED to light up.
b) Why is it essential to connect a resistor in series with the LED?
[4]

10 Write the truth table for the logic circuit shown. [4]

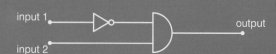

11 Copy and complete the truth table below for each combination of logic gates.

input A	input B	output
0	0	
0	1	
1	0	
1	1	

a)

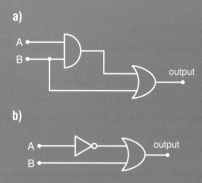

b)

[8]

12 The diagram below shows an alarm system consisting of a window sensor, a key-operated sensor, a logic circuit and an alarm. The window sensor gives logic 0 if the window is open and logic 1 if the window is closed. The key sensor gives logic 1 or 0 according to the position of a switch in the sensor. The system is designed to switch the alarm on (logic state 1) only if the switch is closed (logic state 1) and the window is open (logic state 0).

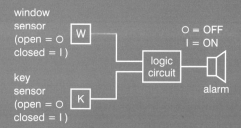

a) Complete the truth table for this system.

window sensor	key sensor	alarm
0	0	
0	1	
1	0	
1	1	

b) Design a logic circuit for this system using an AND gate and a NOT gate. [6]

13 The diagram shows a time-delay circuit consisting of a resistor, a capacitor, a two-pole switch S, an OR gate and a voltmeter.

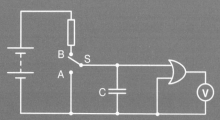

a) When the two–pole switch is at A, what is the reading on the voltmeter?
b) State and explain what happens to the voltmeter reading when this switch is moved from A to B. [5]

14 a) An amplifier has a voltage gain of 200. Calculate the output voltage when the input voltage is 0.020 V.
b) Calculate the output power if the resistance of the output is 10 Ω. [3]

15 An amplifier is connected to a loudspeaker of resistance 8.0 Ω. When a certain input signal is supplied to the amplifier, the voltage across the loudspeaker is 4.0 V. Calculate **a)** the current through the loudspeaker **b)** the power supplied to the loudspeaker. [4]

16 An amplifier is connected to a loudspeaker which has a resistance of 6.0 Ω. When an input signal of voltage 0.10 V is supplied to the amplifier, a voltage of 3.0 V is supplied by the amplifier to the loudspeaker.
a) Calculate **(i)** the voltage gain of the amplifier **(ii)** the output power of the amplifier. [3]
b) The input current is 0.05 A when the input voltage is 0.10 V. Calculate **(i)** the input power supplied to the amplifier **(ii)** the power gain of the amplifier. [3]

Total = 76 marks

Transport

MIND MAP Page 10.

5.1 On the move

preview

At the end of this topic you will be able to:

- describe how to measure average speed and instantaneous speed
- carry out calculations involving the relationship between distance, time and average speed
- define the terms 'speed' and 'acceleration'
- calculate acceleration from change in speed per unit time (mph/s or m/s^2)
- draw speed–time graphs showing steady speed, slowing down and speeding up
- describe the motions represented by speed–time graphs
- calculate acceleration from speed–time graphs for motion with a single constant acceleration
- identify situations where average and instantaneous speeds are different
- explain how the method used to measure the time of travel can have an effect on the measured value of instantaneous speed
- calculate distance moved and acceleration from speed–time graphs for motion involving more than one constant acceleration
- carry out calculations involving the relationship between initial speed, final speed, time and uniform acceleration.

Speed and distance

Fact file
- Speed is defined as distance travelled per unit time.
- The unit of speed is the metre per second (m/s).
- Average speed (in m/s) = distance travelled (in m) / time taken (in s)

Measuring average speed

The average speed of a moving object can be found by recording the time taken for the object to cover a measured distance. The distance can be measured using a metre rule or a measuring tape. The time can be measured with a stopwatch or by using two light gates connected to an electronic timer or a computer.

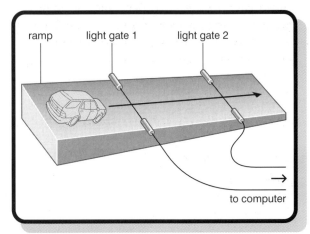

Using a computer to time an object

As the car enters the first light gate, the computer starts timing and it stops timing when the car enters the second light gate. This gives a value for the time taken to travel between the two light gates. The distance between the two light gates is also measured.

The average speed of an object is calculated by dividing the distance travelled by the time taken. The unit of speed is the metre per second, usually abbreviated to m/s. Note that 1 kilometre = 1000 metres. Also, mph stands for miles per hour.

$$\text{Average speed (m/s)} = \frac{\text{distanced travelled (m)}}{\text{time taken (s)}}$$

$$\bar{v} = \frac{d}{t}$$

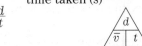

Worked example

A motor cyclist travels a distance of 3 kilometres in a time of 6 minutes. Calculate her average speed in metres per second.

Solution

$$\bar{v} = \frac{d}{t} = \frac{3000 \text{ (m)}}{6 \times 60 \text{ (s)}} = 8.3 \text{ m/s}$$

Measuring instantaneous speed

Instantaneous speed can be found by measuring the average speed over a **very short** time interval. This gives the speed at any instant or particular time and is indicated, for example, by a car speedometer. In the laboratory, the speed of a model car can be found by passing a vehicle fitted with a narrow mask through a light gate connected to a computer connected to an electronic timer. If the width of the mask is measured, then the instantaneous speed is found from:

$$\text{instantaneous speed} = \frac{\text{width of the mask}}{\text{time taken}}$$

The instantaneous speed is in metres per second if the width is converted to metres and divided by the time in seconds.

Question

1 A toy car fitted with a narrow mask of width 4 mm passes through an electronic timer in 0.01 s. Calculate the instantaneous speed.

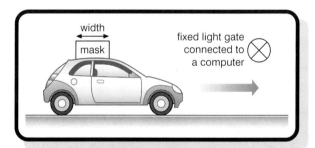

Measuring instantaneous speed

Acceleration

Fact file

Acceleration is defined as the rate of change of speed for an object moving along a straight line.

$$\text{acceleration} = \frac{\text{change of speed}}{\text{time taken for the change}}$$

Answer ↑ 0.4 m/s

Change in speed is measured in m/s or mph, time taken is measured in seconds, so acceleration is measured in metres per second per second (m/s²) or in mph per second (mph/s).

Speed–time graphs

The motion of an object can be described from its speed–time graph. The speed–time graph below shows the motion of a car travelling at constant speed, then accelerating and finally decelerating to a halt.

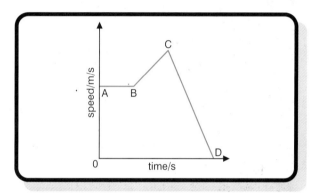

A speed–time graph

From A to B, the car moves at constant speed.

From B to C, the car accelerates uniformly (i.e. it has the same increase in speed every second).

From C to D, the car decelerates uniformly (i.e. it has the same decrease in speed every second).

If the speed–time graph has numerical values on each axis, the size of an acceleration or deceleration may be calculated as shown below.

Worked example

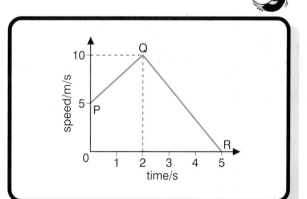

Acceleration from P to Q = change of speed / time taken = 5 (m/s) / 2 (s) = 2.5 m/s²

Acceleration from Q to R = change of speed / time taken = −10 (m/s) / 3 (s) = −3 m/s²

Note: a minus sign is used to represent a negative acceleration because the object decelerates. A negative acceleration is called a deceleration.

Questions

2 A car changes speed from 30 mph to 60 mph in 10 seconds. Calculate the car's acceleration in mph/s.
3 A footballer changes speed from 5 m/s to 8 m/s in 2 s. What is his acceleration?
4 A train increases its speed from zero to 8 m/s in 200 s. Sketch a speed–time graph for this motion and show that the acceleration of the train is 0.040 m/s².

More about speed and acceleration

The difference between average speed and instantaneous speed

When the speed of a moving object changes:

1 its **average speed** for a certain distance is calculated by measuring the time taken to travel that distance using the equation

$$\text{average speed} = \frac{\text{distance}}{\text{time taken}}$$

For example, a bus takes 30 minutes to travel a distance of 18 km. Its average speed for that

$$\text{distance} = \frac{18\,000\,(m)}{30 \times 60\,(s)} = 10 \text{ m/s}.$$

2 its **instantaneous speed** is the speed at any instant and can have a variety of values. In the example above, the instantaneous speed would be indicated on the speedometer of the bus, sometimes below 10 m/s and sometimes above 10 m/s. To measure the speed at any instant, the distance moved in a very short time interval at that instant must be measured. Electronic timers are used to eliminate the problem of human reaction time in manual timing devices.

Acceleration and distance travelled from speed–time graphs

For an object moving along a straight line:

1 $\text{Acceleration} = \dfrac{\text{final speed} - \text{initial speed}}{\text{time taken}}$

This equation may be written in the form:

$$a = \frac{v - u}{t}$$

where v is the final speed in m/s, u is the initial speed in m/s, t is the time taken, and a is acceleration (or deceleration) in m/s².

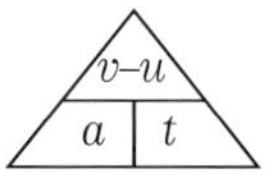

2 The distance moved can be found from the area under a speed–time graph. This is normally done by dividing the total area under the graph into smaller triangles and rectangles.

Worked example

In the graph on page 91, find the distance moved from Q to R.

Solution

The section under QR is a triangle of area $\frac{1}{2} \times 10$ (m/s) $\times 3$ (s).

Therefore the distance moved is 15 m.

Questions

5 For the speed–time graph shown on page 91:
  a) show that the distance moved from P to Q is 15 m
  b) show that the average speed for the whole journey from P to R is 6 m/s
6 An object accelerates at 5 m/s² from 35 m/s to 50 m/s without a change of direction.
  a) Calculate the time taken for this change of speed.
  b) (i) Sketch a speed–time graph to represent this motion.
     (ii) Use the graph to calculate the distance moved and the average speed.
7 The following graph shows how the speed of a train changes as it travels between two stations.
  a) Determine the acceleration and distance moved in each of the three parts of the journey.
  b) Hence calculate the average speed of the train.

Answers
2 3 mph/s
3 1.5 m/s²
4 The graph is a straight line through 0.

Transport

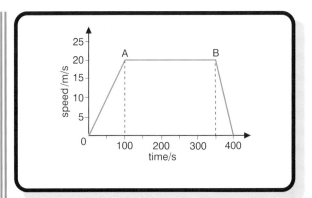

Answers

6 a) 3 s b) (i) The graph is a straight line from 35 m/s to 50 m/s in 3 s. (ii) 127.5 m, 42.5 m/s
7 a) 0 to A: 0.20 m/s²; A to B: 0, 5000 m; B to C: −0.40 m/s², 500 m

5.2 Forces at work

preview

At the end of this topic you will be able to:

- describe the effects of forces in terms of their ability to change the shape, speed and direction of travel of an object
- describe the use of a Newton balance to measure force
- state that weight is a force and is the Earth's pull on an object
- use the approximate value of 10 N/kg to calculate weight
- state that the force of friction can oppose the motion of a body
- describe and explain situations in which attempts are made to increase or decrease the force of friction
- state that equal forces acting in opposite directions on an object are called balanced forces and are equivalent to no force at all
- state that when balanced forces or no forces act on an object, its speed remains the same
- explain, in terms of forces required, why seat belts are used in cars
- describe the effects of change of mass or of force on the acceleration of an object
- carry out calculations involving the relationship between *a*, *F* and *m*
- distinguish between mass and weight
- state that weight per unit mass is called the gravitational field strength
- explain the movement of objects in terms of Newton's First Law of Motion
- carry out calculations involving the relationship between *a*, *F* and *m* involving two or more forces acting on an object in the same direction or in opposite directions.

The effects of forces

Forces can change the shape, speed or direction of an object. For example

1. squeezing a sponge changes its shape
2. kicking a football changes its speed
3. when a car's steering wheel is turned, the car's direction changes.

The unit of force is the **newton (N)**.

A Newton balance is a spring balance calibrated in newtons and used to measure force.

Weight

- The weight of an object is the force of gravity, i.e. the Earth's pull on the object. Weight is measured in newtons.
- The mass of an object is the amount of matter in it. Mass is measured in kilograms.
- All objects near the Earth's surface are pulled towards the Earth's centre with a force of 10 newtons for each kilogram of mass (i.e. 10 N/kg).
- To calculate the weight of an object in newtons, multiply its mass in kilograms by 10.

Using a newton balance

Revise Standard Grade Physics

Friction

Friction is a force which opposes motion and can stop movement completely. Friction is essential in some situations and a nuisance in other situations. For example, the chunky tyres of a mountain bike are designed with deep treads to increase the friction between the tyres and the ground. However, the moving parts of the mountain bike have to be lubricated to reduce the force of friction.

Questions

1 Calculate the weight of a person of mass 55 kg.
2 Explain why cars on icy roads are hard to control and can easily get stuck.

Answers

1 550 N
2 Friction is too low to enable the tyres to grip the road. The motion of the car is difficult to change because there is insufficient grip.

Force and motion

Balanced forces

- A force is anything that can change the speed or direction of an object.
- If there is no force acting on an object, the object moves at constant speed without changing direction, or it remains stationary.
- If an object is acted on by two or more forces which balance each other out, the object either remains at rest or continues to move at constant speed without change of direction. In either situation, the forces are described as **balanced forces**.
- A seat belt in a car is designed to keep a person in the seat in the event of an accident or sudden braking manoeuvre. Without a seat belt, the person would continue to move forward without a change of speed for a short time until he or she collided with the windscreen or the seat in front. Seat belts help to prevent injuries in crashes by providing decelerating forces, which slow down the forward movement of drivers and passengers.

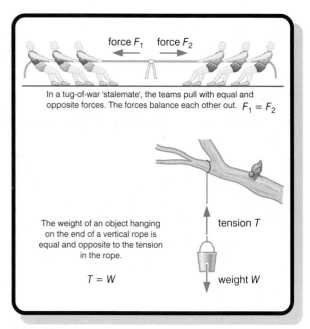

Balanced forces

Acceleration, force and mass

The speed and direction of motion of an object changes if the object is acted on by a force or by several forces which do not balance out. The combined effect of different forces acting on an object is called the **resultant force** on the object. If the resultant force is not zero, the combined effect of the forces is called the **unbalanced force**.

- If an object is at rest or moving at constant speed and direction, the resultant force on it must be zero. This is known as Newton's First Law of Motion.
- If an object's speed or direction of motion is changing, the resultant force on it is not zero. Experiments show that the acceleration of an object is proportional to the unbalanced force on the object. This is known as Newton's Second Law of Motion and it can be written as an equation:

unbalanced force = mass × acceleration
   (in N)   (in kg)   (in m/s²)

$$F_{UN} = ma$$

The **newton** is equal to the force needed to give a 1 kg mass an acceleration of 1 m/s^2.

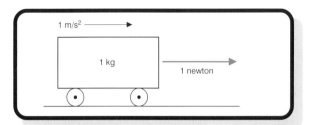

The newton

Maths workshop

Note that $F = ma$ can be rearranged (using the triangle rule below if necessary) as

either $a = \frac{F}{m}$ to calculate a from F and m

or $m = \frac{F}{a}$ to calculate m from F and a

Question

3 An unbalanced force of 40 N is applied to an object of mass of 5 kg. Calculate the resulting acceleration.

Further forces

Mass

The mass of an object is a measure of how much matter an object has. Mass is measured in kilograms (kg).

Weight

The weight of an object is the pull of gravity on it, measured in newtons (N). Gravitational field strength, g, is the weight per unit mass.

$$g = \frac{W}{m}$$

All objects near the Earth's surface are pulled towards the Earth's centre with a force of 10 newtons for each kilogram of mass. So we can say that on the Earth, $g = 10$ N/kg. From the above equation

weight = mass × g
(in N)   (in kg) (in N/kg)

$W = mg$

Hence a 7 kg mass weighs 70 N on Earth
a 2.8 kg mass weighs 28 N on Earth
g is different on different planets. For example, $g = 3.7$ N/kg on Mars so a 7 kg mass on Mars weighs 26 N.

Answer 3 8 m/s²

Transport

Unbalanced forces and Newton's Second Law

To calculate the acceleration of an object acted on by two forces in opposite directions:

1 calculate the unbalanced force F_{UN} from the difference of the two forces
2 calculate the acceleration using Newton's Second Law in the form $a = \frac{F_{UN}}{m}$

Questions

4 A motor cyclist and motorbike have a combined mass of 220 kg. The engine exerts a force of 1600 N and the frictional forces opposing the motion total 280 N. Calculate the acceleration of the motorcyclist.

5 A car of total mass 800 kg can accelerate from rest to a speed of 20 m/s in 20 s. Calculate
  a) the acceleration of the car
  b) the force producing this acceleration, assuming drag forces are negligible
  c) the ratio of this force to the total weight of the car. Assume $g = 10$ m/s²

Answers 4 6 m/s² 5 a) 1 m/s² b) 800 N c) 0.1

5.3 Movement means energy

preview

At the end of this topic you will be able to:

- describe the main energy transformations as a vehicle accelerates, moves at constant speed, brakes and goes up or down a slope
- state that work done is a measure of the energy transferred
- carry out calculations involving the relationship between work done, force and distance
- carry out calculations involving the relationship between power, work and time
- state that the change in gravitational potential energy is the work done against or by gravity
- state that the greater the mass or the speed of a moving object, the greater is its kinetic energy

- carry out calculations involving the relationship between kinetic energy, mass and speed
- carry out calculations involving energy, work, power and the principle of conservation of energy.

Energy transformations

★ **Work is done when a force makes an object move**. The greater the force or the further the movement, the greater the amount of work done. For example, the work done to lift a box by a height of 2 metres is twice the work done to lift the same box by a height of 1 metre.

★ **Energy is the capacity to do work**. A battery-operated electric motor used to raise a weight does work on the weight when it is raised. The battery therefore contains energy because it has the capacity to make the motor do work.

★ **Heat is energy transferred due to a difference of temperature**. Heat and work are two methods by which energy can be transferred to or from a body. Work is energy transferred due to force and heat is energy transferred due to temperature difference. Heat as a method of transferring energy is discussed in more detail on page 111.

★ **Energy can be transferred from one object to other objects**. A raised weight has the capacity to do work and therefore contains energy. For example, it could be used to keep a clock running.

Forms of energy
Energy exists in different forms, including:
- **kinetic energy**, which is the energy of a moving body due to its motion
- **potential energy**, which is the energy of a body due to its position
- **chemical energy**, which is energy released due to chemical reactions
- **light energy**, which is energy carried by light
- **elastic energy**, which is energy stored in an object by changing its shape
- **electrical energy**, which is energy due to electric charge
- **nuclear energy**, which is energy released due to nuclear reactions
- **sound energy**, which is energy carried by sound waves
- **heat energy**, which is the energy of an object due to its temperature.

An energetic journey
During a long car journey, the chemical energy provided by the fuel will be changed or transformed into several different forms of energy (e.g. kinetic energy, heat, etc.). In the diagram below, a car accelerates from rest, goes up and then down a hill, then brakes to a halt.

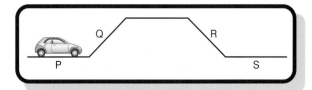

Throughout the journey, wasted energy in the form of **sound** and **heat** is produced.

P: accelerating from rest, chemical energy → kinetic energy

Q: going up a hill, kinetic energy → gravitational potential energy

R: going down a hill, gravitational potential energy → kinetic energy

S: braking to a halt, kinetic energy → heat

Measuring energy
Work and energy are measured in **joules (J)**, where 1 joule is defined as the work done when a force of 1 newton acts over a distance of 1 metre in the direction of the force.

The following equation is used to calculate the work done or energy transferred by a force:

work done = force × distance moved in the
(or energy   direction of the force
transferred)

E_w = F × d
(in J) (in N) (in m)

Transport

Question

1 a) A student applies a force of 600 N to a wardrobe and pushes it a distance of 2 m across a floor. Calculate the work done by the student.

b) If the wardrobe had been empty, it could have been pushed across the floor using a force of 200 N. How far could it have been moved for the same amount of work?

Gravitational potential energy

If a mass m is moved through a height h, its change of gravitational potential energy is $E_p = mgh$, where g is the acceleration due to gravity.

This is because the force of gravity on the object (i.e. its weight) is mg.

1. Any object at rest is acted on by a support force which is equal and opposite to its weight.
2. If raised, the work done by the support force is mgh (equal to force × distance). This is therefore the gain of potential energy of the object.
3. If lowered, the work done by gravity is mgh (equal to force × distance). This is therefore the decrease in potential energy of the object.

Note: for an object of mass m, its weight = mg because g is the force of gravity per unit mass on an object. Also, the mass must be in kilograms and the height gain in metres to give the gain of potential energy in joules. The value of g on the Earth is 10 N/kg.

Questions

2 How much gravitational potential energy does a 60 kg swimmer gain as a result of climbing a height of 15 m?

3 If the swimmer in question 2 dives from a diving board 15 m above the water, how much kinetic energy does the swimmer have just before impact?

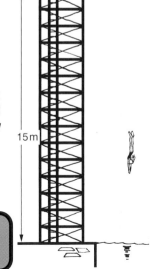

Answers

1 a) 1200 J b) 6 m   2 9000 J   3 9000 J

A very important principle

The **Principle of Conservation of Energy** states that in any change, **the total energy before the change is equal to the total energy after the change**. In other words, the total energy is conserved even though it may change from its initial form into other forms as a result of the change. Consider the energy changes of a 100 N weight released at a height of 1.0 m above the floor.

- Its initial potential energy (relative to the floor) = mgh = 100 N × 1.0 m = 100 J
- Its potential energy (relative to the floor) just before impact = 0 since its height is effectively zero just before impact.

It therefore loses 100 J of potential energy as a result of falling to the floor. Just before impact, its kinetic energy is therefore 100 J since it has no kinetic energy at the start and all its potential energy is transformed into kinetic energy – assuming no air resistance!

But

The trouble with energy is that it tends to spread out when it changes from one form into other forms. For example, in a bicycle freewheeling down a slope, friction at the wheel bearings will probably cause the bearings to become warm. Some of the initial potential energy is therefore converted into heat energy. Such heat energy is lost to the surroundings and can never be recovered and used to do work. The heat energy is therefore wasted. Even where energy is concentrated, such as when a car battery is charged, energy is wasted in the process. For example, the electric current passing through a car battery when it is being charged will warm the circuit wires a little. You can't win with energy!

Power

Power is defined as the work done or energy transferred per second.

power (in W) = energy transferred (in J) / time taken (in s)

$P = \dfrac{E}{t}$

Revise Standard Grade Physics

The unit of power is the **watt** (W), equal to 1 joule per second. Note that 1 kilowatt (kW) equals 1000 watts and 1 megawatt (MW) equals 1 million watts.

Question

4 Calculate the muscle power of a student of mass 50 kg who climbs a height of 5.0 m up a rope in 10 seconds. (*Hint:* remember that weight = mass × g, where g = 10 N/kg. Also, the student uses both arms to climb the rope.)

Answer

4 125 W

On the road

Kinetic energy

For a mass m moving at speed v, its kinetic energy = $\frac{1}{2} mv^2$

$$E_k = \frac{1}{2} mv^2$$

To use this formula, mass must be in kilograms and speed in metres per second. The kinetic energy is then in joules. For example, the kinetic energy of a ball of mass 2 kg moving at a speed of 3 m/s is $\frac{1}{2} \times 2 \times 3^2 = 9\,J$.

Practice questions

Set out your answers clearly. Where necessary assume $g = 10\,m/s^2$.

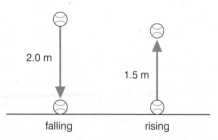
falling   rising

1 A tennis ball of mass 0.20 kg is released from rest at a height of 2.0 m above a concrete floor. It rebounds to a height of 1.5 m. Calculate
 a) its kinetic energy and speed just before impact
 b) its kinetic energy and speed just after impact
 c) its loss of energy between release and its maximum height after rebounding.

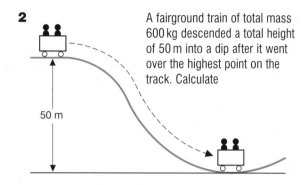

2 A fairground train of total mass 600 kg descended a total height of 50 m into a dip after it went over the highest point on the track. Calculate

 a) its loss of potential energy in this descent
 b) its kinetic energy and speed at the bottom of the dip, assuming air resistance was negligible and its kinetic energy at the highest point was zero.

3 An aeroplane of mass 600 kg takes off from rest in 50 s over a distance of 1500 m.
 a) Calculate (i) its speed when it lifts off
   (ii) its acceleration during take-off
   (iii) the force needed to produce this acceleration.
 b) Why is the engine force greater than the force calculated in **a) (iii)**?

4 A hot air balloon and its occupants have a total weight of 5000 N. It is descending at a constant speed of 0.5 m/s.

Constant speed

 a) What is the total upward force on it during this descent?
 b) Calculate its kinetic energy.
 c) Calculate its loss of height and loss of potential energy in 1 minute.
 d) Explain why it does not gain kinetic energy.

Answers

1 a) 4 J, 6.3 m/s b) 3 J, 5.5 m/s c) 1 J
2 a) 300 kJ b) 300 kJ, 32 m/s
3 a) (i) 60 m/s (ii) 1.2 m/s² (iii) 720 N
 b) Because of air resistance.
4 a) 5000 N b) 62.5 J c) 30 m, 150 kJ d) All the potential energy is transferred to the air by the upward force.

Transport

Equations review

1 $\overline{v} = \dfrac{d}{t}$

$\overline{v}$ is the average speed in m/s, d is the distance moved in m and t is the time taken in s.

Note: the instantaneous speed $v = \dfrac{d}{t}$ is measured over a small time interval.

2 $a = \dfrac{v - u}{t}$

a is the acceleration in m/s², u is the initial speed in m/s, and v is the speed in m/s after time t in s.

3 $W = mg$

W is the weight in N of an object of mass m in kg and g is the gravitational field strength (in N/kg). On the Earth's surface, $g = 10$ N/kg.

4 $F_{UN} = ma$

F_{UN} is the unbalanced force in N acting on a mass m in kg, and a is the acceleration in m/s².

5 $E_w = Fd$

E_w is the work done in joules by a force F in N acting over a distance d in m in the direction of the force.

6 $E_p = mgh$

E_p is the change of potential energy in J when a mass m in kg is raised or lowered through a height h in m.

7 $E_k = \dfrac{1}{2} mv^2$

E_k is the kinetic energy in J of a mass m in kg moving at speed v in m/s.

round-up

The acceleration of a freely falling object $g = 10$ m/s².

1 A walker travelled a distance of 10 km in 2 hours. Calculate the walker's average speed in **a)** km/h **b)** m/s. [2]

2 a) A walker leaves a car park and walks at a steady speed of 1.2 m/s for 1 hour. How far did the walker travel in this time in **(i)** metres **(ii)** kilometres? [2]
b) A runner leaves the same car park 40 minutes after the walker and catches up with the walker at a distance of 4 km from the car park. What was the runner's speed? [1]

3 a) What feature of a graph of speed against time gives the distance travelled? [1]
b) Which section of the graph below shows **(i)** constant speed **(ii)** constant rate of increase of speed? [2]

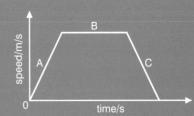

4 A car is travelling at a speed of 15 m/s when the driver sees a tree lying across the road ahead and is forced to brake. The graph shows how its speed changed with time.

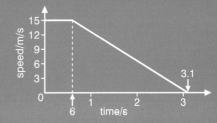

a) If the driver's reaction time is 0.6 s, how far does the car travel before the driver applies the brakes? [1]
b) The car takes 2.5 s to stop after the brakes are first applied. Calculate its deceleration. [1]
c) (i) Use the graph to calculate the braking distance. [1]
(ii) Hence calculate the total stopping distance. [1]

5 The graph shows the progress of two cyclists in a 10 km road race.
a) One cyclist X maintained a constant speed throughout. What can you deduce from the graph about the speed of the other cyclist?

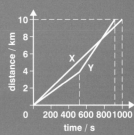

[4]
b) (i) From the graph, calculate the speed of X. [1]
(ii) From the graph, calculate the speed of Y when Y overtook X. [1]

6 a) Explain the difference between instantaneous speed and average speed. [2]

b) A police car joins a motorway and travels north at a constant speed of 30 m/s for 5 minutes. It then leaves the motorway at a motorway junction, rejoins it immediately and travels south in the opposite direction for 20 minutes at a steady speed of 20 m/s to the scene of an accident.

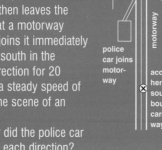

(i) How far did the police car travel in each direction? [2]

(ii) How far from the point where the police car first joined the motorway was the scene of the accident? [1]

7 A 1000 kg car accelerates from rest to a speed of 10 m/s in 20 s. Calculate the acceleration of the car and the force needed to produce this acceleration. [2]

8 Why does a parachutist fall at constant speed? [3]

9 Explain why the shape of a vehicle affects the top speed. [3]

10 A vehicle of mass 700 kg accelerates steadily from rest to a speed of 8 m/s in a time of 5 s. The graph shows how the speed of the vehicle increased with time.

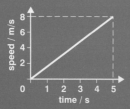

a) Calculate the acceleration of the vehicle and the distance travelled in this time. [2]

b) What force acting on the vehicle is necessary to produce this acceleration? [1]

11 A train of total mass 30 000 kg is travelling at a constant speed of 10 m/s when its brakes are applied, bringing it to rest with a constant deceleration in 50 s.

a) Sketch a graph to show how the speed of the train changed with time. [2]

b) Calculate (i) the distance moved by the train in this time (ii) its acceleration. [2]

c) Hence calculate the force of the brakes on the train. [1]

d) Calculate the ratio of the braking force to the train's weight. [1]

12 a) Explain why a moving car skids if the brakes are applied too hard. [1]

b) Explain why the stopping distance of a car travelling at a certain speed is greater if the road surface is wet. [4]

c) The graph shows how the speed of a car of mass 1200 kg on a dry road decreases with time when it stops safely in the shortest possible distance. The car is initially moving at a speed of 15 m/s.

Calculate
(i) the braking distance
(ii) the deceleration of the car and the braking force. [3]

d) On a wet road, the braking force is reduced by half. Calculate the time it would take to brake safely from a speed of 15 m/s on a wet road and determine the braking distance in this condition. [2]

13 a) Fill in the spaces:
In any energy transformation, energy cannot be _____ or _____ . This means energy is_____ . [3]

b) A crate of mass 60 kg is lifted 1.8 m on to the back of a truck. Calculate the gravitational potential energy gained by the crate. [2]

14 An aeroplane of total mass 8000 kg is in level flight at a constant speed of 60 m/s.

a) Calculate its weight and the lift force acting on the aeroplane. [2]

b) The output power of its engines is 200 kW. Calculate

(i) the distance it moves in 100 s

(ii) the energy output from the engine in this time. [2]

c) What happens to the energy supplied by the engine when the aeroplane is moving at constant speed in level flight? [1]

15 A lift of total mass 400 kg descends at a constant speed of 2 m/s.

a) Calculate

(i) its weight and its kinetic energy

(ii) its loss of potential energy per second. [3]

b) A What happens to the potential energy lost by the lift? [3]

16 a) An athlete runs a distance of 100 m in a time of 10.5 s. Calculate the athlete's average speed over this distance [1]

b) The athlete's mass is 60 kg. Calculate the athlete's kinetic energy at the speed calculated in **a)**. [1]

c) If all the kinetic energy calculated in **b)** could be converted into potential energy by the athlete, what height gain would be possible? [1]

d) A pole vaulter is capable of jumping considerably higher. Discuss how this is achieved. [1]

17 a) A ball of mass 0.2 kg is thrown directly upwards with an initial speed of 25 m/s. Calculate how long it took to reach its highest point, assuming $g = 10$ m/s^2. [1]

b) The graph shows how its vertical position changed after it left the thrower's hand.

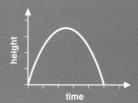

(i) What was its initial kinetic energy? [1]

(ii) Calculate its maximum gain of potential energy. [1]

(iii) Calculate its maximum gain of height. [1]

Total = 74 marks

Energy matters

6.1 Supply and demand

MIND MAP Page 10.

preview

At the end of this topic you will be able to:
- state that fossil fuels are at present the main sources of energy
- state that the reserves of fossil fuels are finite
- explain one means of conserving energy related to the use of energy in industry, in the home and in transport
- carry out calculations relating to energy supply and demand
- classify renewable and non-renewable sources of energy
- explain the advantages and disadvantages associated with at least three renewable energy sources.

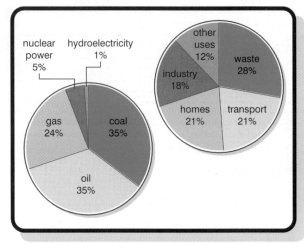

UK energy supplies   UK energy demands

The total energy demand for the United Kingdom in one year is about 10 million million million joules. This works out at about 5000 joules per second for every person in the country. The pie charts show how the total energy is obtained and how it is used.

Fuels are substances which release energy as a result of changing into another substance. Fuels cannot be reused.

- fossil fuels – coals, oil, gas, wood
- nuclear fuels – uranium, plutonium

Our main sources of energy at present are **coal**, **oil** and **gas**, which are classified as **fossil fuels**. It has been estimated that oil and gas will last about 50 years and coal reserves will be exhausted in about 300 years if the present rate of consumption continues. There is only a limited amount of the three fossil fuels left, so we say that our reserves are **finite**.

Renewable energy resources are sources of useful energy which do not change the substances involved, allowing them to be reused. The energy usually comes from the Sun.

- solar-driven resources – solar panels for water heating, solar cells
- weather-driven resources (indirectly powered by the Sun's heating effect on the atmosphere) – hydroelectricity, aerogenerators (wind), wave powered generators
- tidal generators (powered by the gravitational potential energy between the Earth and the Moon)
- geothermal power (powered by heat energy in the Earth's interior)

Questions

1. Giant windmill dynamos called aerogenerators can produce 300 W of power for each square metre of blade if the wind is blowing at 10 m/s. Calculate the blade area required to produce an output of 600 kW if the wind speed is 10 m/s.

2. How many such aerogenerators would be needed to give a total output of 30 MW?

Answers
1. Area = 600 000 W/300 W = 2 000 m²
2. Number required = 30 × 10⁶W/600 × 10³W = 50

Checklist of sources of energy

renewable	non-renewable
waves, tidal	coal
solar, wind	oil
hydroelectric	gas
geothermal	nuclear fuel

Energy efficiency

Reasons why energy should not be wasted:

1. Fuel supplies are finite and cannot be renewed once used.
2. Fossil fuels release carbon dioxide gas which is thought to be causing global warming, resulting in melting icecaps and rising sea levels. Sulphur dioxide from power stations is a cause of acid rain.
3. Nuclear fuel creates radioactive waste which must be stored safely for hundreds of years to prevent it harming us.
4. Small-scale renewable resources may damage the environment, for example turbine noise from aerogenerators and effects on plant and animal life when tidal power stations are made.
5. It costs money to make energy useful and to distribute it.

How to use energy resources more efficiently

| industry | Use 'waste' heat to produce low cost heating for homes, offices and schools |
	Improve efficiency of manufacturing techniques
home	Insulate houses
	Turn down thermostats
	Switch off all unnecessary lights and get central heating checked regularly
transport	Car sharing
	Make more use of public transport
	Drive more slowly
	Use cars with smaller, more efficient engines and get them serviced more regularly

Renewable energy sources: advantages and disadvantages

renewable energy source	advantages	disadvantages
hydroelectric	No pollution	Expensive to set up
	Energy is free	Large areas of land are flooded
	Can produce large amounts of electricity	Only possible in certain places
solar	No pollution	Expensive to install
	Energy is free	Sunny climate required
		Large areas need to be covered with solar panels
wind	No pollution	Noisy
	Energy is free	Only produces energy when it is windy
		Expensive to install
		Large numbers of wind generators required to give reasonable amounts of energy

6.2 Generation of electricity

preview

At the end of this topic you will be able to:
- **identify from a diagram the energy transformation at each stage of:**
  - **– a thermal power station**
  - **– a hydroelectric power station**
  - **– a nuclear power station**
- **state that radioactive waste is produced by a nuclear reactor**
- **carry out calculations on energy transformation to include gravitational potential energy**

- describe the principle and give the advantages of a pumped hydroelectric scheme
- compare energy output from equal masses of coal and nuclear fuels
- carry out calculations involving efficiency of energy transformations
- state that energy is degraded in energy transformations
- explain a chain reaction in simple terms.

Energy transformations in power stations

Thermal power station (fuel – coal, oil or gas)

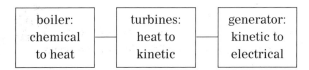

Nuclear power station (fuel – uranium)

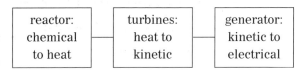

The nuclear reactor is at the heart of a nuclear power station and unfortunately produces radioactive waste which can remain radioactive for thousands of years. The waste is stored for long periods until it is much safer.

Hydroelectric power station (fuel – water)

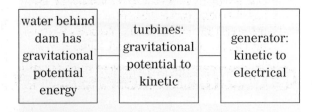

Hydroelectric power station

Water stored behind a dam stores gravitational potential energy E_p, which can be calculated from:

$E_p = mgh$

The gravitational potential energy is converted into kinetic energy E_k and eventually into electrical energy E_e, which can be calculated from:

$E_e = Pt$

Worked example

A small hydroelectric power station allows water to fall vertically through 25 m. If the average mass of water falling per second is 50 000 kg, calculate the power output of the water flow.

Solution

Gravitational potential energy of the water,

$E_p = mgh$

$= 50\,000 \times 10 \times 25$

$= 12\,500\,000\,\text{J}$

Power output of water flow, $P = E_p/t$

$= 12\,500\,000/1$

$= 12\,500\,000\,\text{W}$

A pumped storage hydroelectric scheme

The demand for electricity varies greatly over a 24-hour period. One method used by electricity companies to cope with this is to use off-peak electricity from their power stations. Electricity produced during the night is used to run turbines in reverse to pump water from a lower reservoir to an upper reservoir.

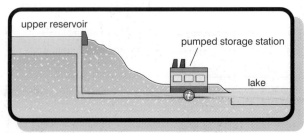

A pumped storage station. Electricity is used to pump water uphill when the demand for electricity is low. Electricity is generated by allowing water to flow downhill when demand is high.

When there is a huge demand, water is allowed to run back down to the lower reservoir producing electricity. This can be done very quickly in response to a sudden demand.

Energy matters

Question

1 Water flows through a hydroelectric power station at a rate of 10 000 kg/s after falling through a vertical height of 50 m. Calculate the gravitational potential energy released by the water each second. Assume the weight of 1 kg is 10 N.

Energy transformations

Energy cannot be created or destroyed, but when energy is transformed from one form to another, there is always less **useful** energy produced. Some of the original energy changes to heat and sound. Energy is degraded, i.e. the quantity is the same but the quality is inferior.

Machines at work

A machine is a device designed to do work. The diagram shows some examples of machines at work. In each case, energy is supplied to the machine enabling it to do **useful work**. The useful work done by a machine is always less than the energy supplied to it because of friction between its moving parts. This causes heating and therefore wastes energy. Energy may be wasted in other ways in a machine as well as through friction.

Machines at work

Efficiency

The efficiency of a machine is the proportion of the energy supplied to the machine which is transferred to useful work. This may be expressed as an equation:

$$\text{efficiency} = \frac{\text{useful work done by the machine}}{\text{energy supplied to the machine}}$$

Notes

1. *Efficiency is sometimes expressed as a percentage (the fraction above multiplied by 100).*
2. *The efficiency of any machine is always less than 100% because of friction.*
3. *Efficiency may be expressed in terms of power as*

$$\frac{\text{power output}}{\text{power input}}$$

Worked example

A pumped storage scheme is 80% efficient. If the input power is 8 MW, calculate the power produced.

Solution

% efficiency = (power output/power input) × 100

80 = [power output/(8 × 10^6)] × 100

power output = (80 × 8 × 10^6)/100

= 6.4 MW

Question

2 A pulley system is used to raise a crate of weight 500 N through a height of 2.4 m. To do this, the operator must pull on the rope with a force of 300 N through a distance of 4.8 m.

a) Calculate **(i)** the work done by the operator **(ii)** the potential energy gain of the crate.

b) Hence calculate the efficiency of the pulley system and give two reasons why it is not 100%.

Answer 2 a) (i) 1440 J (ii) 1200 J b) 83%

Answer 1 5 MJ/s

Releasing energy from the nucleus

★ The uranium-235 nucleus is unstable and splits into two approximately equal 'daughter' nuclei, and two or three neutrons. This splitting process is called **fission**.

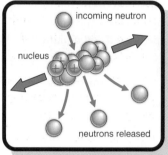
Fission

★ Energy is released when a U-235 nucleus splits, which is carried away as kinetic energy by the daughter nuclei and neutrons.

★ Fission can be induced by bombarding U-235 nuclei with neutrons.

★ A controlled **chain reaction** is created if there are sufficient U-235 nuclei. One neutron from each fission event goes on to cause the fission of another nucleus.

★ To keep control of the reactor, some neutrons are absorbed to avoid too many fissions. If they were not absorbed, the reactor would go out of control. When the original uranium atoms split, heat energy is produced and this is used to produce steam to drive turbines.

Coal versus nuclear fuel

Uranium is the fuel used in the reactors of nuclear power stations. Coal is still widely used as the fuel in a thermal power station. It is interesting to compare the energy output of 1 kg of each fuel.

1 kg of uranium produces 3×10^{12} J of energy

1 kg of coal produces 3×10^{6} J of energy.

In other words uranium produces about 1 million times more energy for the same mass of fuel!

Worked example

Each kilogram of coal burned in a thermal power station produces 3×10^{6} J of heat and light energy. If the efficiency of the power station is 35%, what is the electrical energy produced by each kilogram of coal?

Solution

% efficiency = (useful E_{output}/total E_{input}) × 100

$35 = [\text{useful } E_{output}/(3 \times 10^{6})] \times 100$

useful $E_{output} = (35 \times 3 \times 10^{6})/100$

$= 1.05 \times 10^{6}$ J

Question

3 A nuclear power station produces 1000 MW of electrical power at an efficiency of 40%.
a) Calculate the energy released per second by the uranium fuel in this power station.
b) Each kilogram of the uranium fuel in this power station releases 3×10^{12} J. Calculate the mass of uranium fuel used in (i) 1 second (ii) 1 day.

Answer

3 a) 2500 MW b) (i) 8.3×10^{-4} kg (ii) 7.2 kg

6.3 Source to consumer

preview

At the end of this topic you will be able to:
- identify circumstances in which a voltage will be induced in a conductor
- identify on a given diagram the main parts of an a.c. generator
- state that transformers are used to change the magnitude of an a.c. voltage
- describe the structure of a transformer
- carry out calculations involving the relationship between V_s, V_p, n_s and n_p
- state that high voltages are used in the transmission of electricity to reduce power loss
- describe quantitatively the transmission of electrical energy by the National Grid system
- explain, using a diagram, how an a.c. generator works

Energy matters

- state the main differences between a full-size generator and a simple working model
- state the factors which affect the size of the induced voltage, i.e. field strength, number of turns on the coil, relative speed of magnet and coil
- explain why a transformer is not 100% efficient
- carry out calculations on transformers involving input and output voltages, turns ratio, primary and secondary currents and efficiency
- carry out calculations involving power loss in transmission lines.

★ If a coil of wire surrounding a magnet is moved left or right, a voltage is induced in the coil. If the magnet is moved towards or away from the coil, again a voltage is induced. The kinetic energy of the moving coil or magnet is transformed into electrical energy.

★ If the wire is part of a complete circuit in which there is no other voltage source, the induced voltage drives a current round the circuit.

★ The faster the wire moves across the field lines, the greater the induced voltage.

★ The stronger the magnetic field, the greater the induced voltage.

A simple a.c. generator

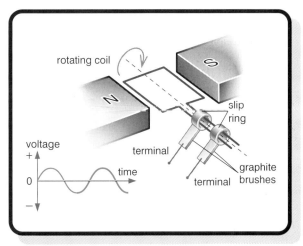

An a.c. generator

Induced voltage

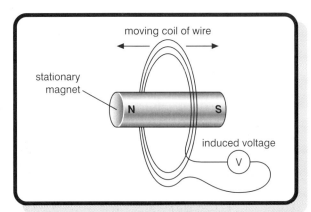

Coil moving, magnet stationary

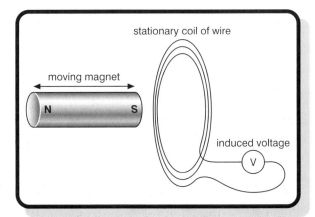

Coil stationary, magnet moving

This device is used to produce electricity. It is really the reverse of an electric motor, i.e. an electric motor changes electrical energy into kinetic energy but an a.c. generator changes kinetic energy into electrical energy.

The essential parts of an a.c. generator are:

- the rotating coil
- the magnet
- the slip rings
- the carbon brushes – these are made of graphite, which is a good conductor, makes good contact on the rings and has low friction.

Revise Standard Grade Physics

Question

1 Why are the brushes of an a.c. generator made from carbon in the form of graphite?

Transformers

The mains voltage is 230 V, 50 Hz a.c. However, many household appliances require voltages other than 230 V to operate correctly. Transformers can change the mains voltage to the desired voltage.

A transformer consists of a soft iron core with two insulated coils of wire called the **primary** and **secondary coils** wound on to it.

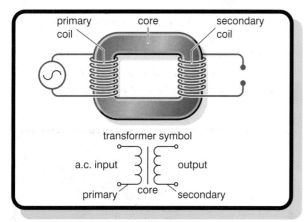

A model transformer

A transformer steps an alternating voltage up or down.

1. The voltage to be transformed is applied to the primary coil of the transformer.
2. The alternating current through the windings of the primary coil creates an alternating magnetic field in the transformer's iron core.
3. The continuously changing magnetic field through the core induces an alternating voltage in the secondary coil.

There are basically two types of transformer, namely **step-up** and **step-down** transformers. A step-up transformer increases the size of the output voltage whereas a step-down transformer decreases the size of the output voltage. These effects are achieved by altering the ratio of the number of turns of wire in the primary and secondary coils.

The transformer rule

If n_p = number of turns in the primary coil and n_s = number of turns in the secondary coil,

if $n_s > n_p$ we have a step-up transformer

if $n_s < n_p$ we have a step-down transformer

n_p/n_s is called the 'turns ratio'

V_p/V_s is called the 'voltage ratio'

$$n_p/n_s = V_p/V_s$$

Worked example

A step-down transformer has 100 turns in the secondary coil and 2000 turns in the primary coil. If the voltage in the primary coil is 230 V, what is the voltage in the secondary coil?

Solution

$n_p/n_s = V_p/V_s$

$2000/100 = 230/V_s$

$V_s = (100 \times 230)/2000 = 11.5 \text{ V}$

Worked example

A step-up transformer has 800 turns in the primary coil and 2000 turns in the secondary coil. If the secondary voltage is 150 V, what will be the voltage in the primary coil?

Solution

$n_p/n_s = V_p/V_s$

$800/200 = V_p/150$

$V_p = 60 \text{ V}$

Question

2 A transformer has a primary coil consisting of 200 turns and a secondary coil consisting of 4000 turns.
   a) Is this transformer a step-up or a step-down transformer?
   b) An alternating voltage of 10 V is applied to the primary coil. Calculate the output voltage across the terminals of the secondary coil.

Answer

1 Graphite is a conductor and it also keeps good contact with the split rings without too much friction.

Answer

2 a) Step-up b) 200 V

Transmission of electricity

Electricity is transmitted over long distances through overhead cables supported by pylons. Energy is lost due to current producing heat in the cables. To cut down the energy losses, the current has to be reduced. By transmitting the electricity at high voltages, the current is automatically reduced (since the power output is $V \times I$, if $V\uparrow$ then $I\downarrow$).

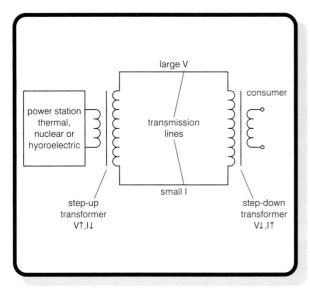

Transmission lines

The National Grid

This is a network of pylons and cables which transmits the electricity produced in power stations to consumers all over the country. Very high voltages (such as 400 kV) are used to reduce power loss in the cables by reducing the current. A step-up transformer is used at the power station end and a step-down transformer at various substations, which reduce the voltage to 33 kV for heavy industry, 11 kV for light industry and 230 V for domestic use.

How an a.c. generator works

The rotating coil of wire cuts through the magnetic field between the magnets. This induces a voltage across the ends of the coil where the slip rings are located. Carbon brushes are in contact with the slip rings and take the output from the generator. The size of the induced voltage output can be increased if:

- stronger magnets are used, increasing the magnetic field strength
- the number of coils of wire is increased
- the coils are rotated more quickly.

Commercial generators

Generators used in power stations are very large. They can produce up to 500 MW, i.e. 20 000 A at 25 kV. They have several differences from the simple generator described above:

- the permanent magnets are replaced by electromagnets
- the single coil is replaced by several fixed coils
- the magnetic field is produced by the rotor, which rotates electromagnets
- the alternating voltage is produced in the stator, which consists of several stationary coils.

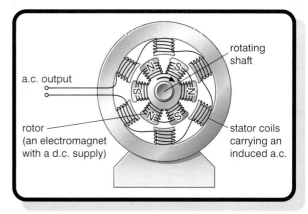

Commercial generator

Transformer efficiency

The percentage efficiency of a transformer is defined as

$$\frac{\text{power output}}{\text{power input}} \times 100\%$$

Transformers are less than 100% efficient because of energy losses due to:

- heat generated in the coils due to the flow of current
- heat generated in the core by currents induced in the core by the changing magnetic field

- energy transformed in the core due to the fact that it is continually being magnetised and demagnetised.

This means that power output < power input. Since $P = IV$, then

$V_s \times I_s < V_p \times I_p$

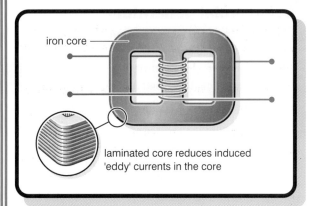

A practical transformer

Note that in an ideal transformer, i.e. one which is 100% efficient, the power in the primary coil will be equal to the power in the secondary coil.

primary power = secondary power

$V_p \times I_p = V_s \times I_s$

$V_p/V_s = I_s/I_p$

thus we have the transformer equation:

$$n_p/n_s = V_p/V_s = I_s/I_p$$

Worked example
A step-down transformer has 6000 turns in its primary coil and 200 turns in its secondary coil. If the primary current is 0.2 A what will be the current in the secondary coil? (Assume 100% efficiency.)

Solution
$n_p/n_s = I_s/I_p$

$6000/200 = I_s/0.2$

$I_s = 6$ A

Worked example
A transformer has a power input of 5 kW. The secondary voltage is 450 V and the secondary current is 10 A. Calculate its efficiency.

Solution
% efficiency = (power output/power input) × 100

$= (V_s \times I_s/5000) \times 100$

$= [(450 \times 10)/5000] \times 100$

$= 90\%$

Question

3 A transformer has a 100-turn primary coil and a 2000-turn secondary coil. A 240 V, 60 W lamp is connected to the secondary coil. Calculate **a)** the primary voltage needed to make the lamp light normally **b)** the primary current when the lamp lights normally, assuming the transformer is 100% efficient.

Power loss in transmission lines

Transmitting power at high voltage and low current is very efficient, with only about 3% loss in electrical energy.

Power loss is calculated using

$P = I^2R$

Worked example
Calculate the power loss in transmission lines of total resistance 15 Ω if the current flowing is 250 A.

Solution
Power loss $= I^2R = (250)^2 \times 15$

$= 937\,500$ W

Worked example
4.5 MW of power is lost in a transmission line with a total resistance of 25 Ω. Find the current in the transmission line.

Solution
Power loss $P = I^2R$

$I^2 = P/R = (4.5 \times 10^6)/25$

$I^2 = 180\,000$

$I = \sqrt{180\,000} = 424.3$ A

Question

4 A transmission line of resistance 10 Ω is used to transfer 400 MW of electrical power at a voltage of 400 kV. Calculate: **a)** the current in the transmission line **b)** the power wasted due to the resistance of the power line.

6.4 Heat in the home

Answers

3 a) 12V b) 5A
4 a) 1000A b) 10MW

preview

At the end of this topic you will be able to:

- use the following terms correctly in context: 'temperature', 'heat', and 'degrees Celsius'
- describe two ways of reducing heat loss in the home due to conduction, convection and radiation
- state that heat loss in a given time depends on the temperature difference between the inside and the outside of the house
- state that the same mass of different materials requires different quantities of energy to raise the temperature of a unit of mass by one degree
- carry out calculations based on practical applications involving heat, mass, specific heat capacity and temperature change
- give examples of applications which involve a change of state, e.g. refrigerator or picnic box cooler
- use the following terms correctly in context: 'specific heat capacity', 'change of state', 'latent heat of fusion' and 'latent heat of vaporisation'
- state that a change of state does not involve a change of temperature
- state that energy is gained or lost by a substance when its state is changed
- use the principle of conservation of energy to carry out calculations on energy transformations which involve temperature change, e.g. $ItV = E_h = cm\Delta T$
- carry out calculations involving specific latent heat.

Temperature, heat and degrees Celsius

Temperature (T) is a measure of how hot or cold an object is.

Heat (E_H) is a form of energy measured in joules (J), and is transferred between objects at different temperatures.

Degrees Celsius (°C) is the unit in which temperature is measured, e.g. ice melts at 0°C and water boils at 100°C.

Heat energy can move from one place to another in three distinct ways:

Conduction	Convection	Radiation
heat energy is passed along solids by vibrating molecules	liquids and gases become less dense and move upwards when heated	infrared waves produced by any hot object and can move through a vacuum

Heat conduction – five facts

1. Heat conduction is heat transfer through a substance without the substance moving.
2. Solids, liquids and gases all conduct heat.
3. Good heat conductors are also good electrical conductors. They contain free electrons which can transport both energy and charge through the substance.
4. Metals and alloys are the best conductors of heat.
5. Insulating materials such as wood, fibreglass and air are very poor conductors of heat. The presence of air pockets in an insulating material improves its insulating properties.

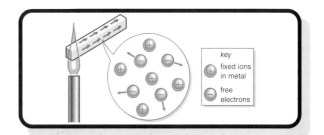

Heat conduction in a metal

Revise Standard Grade Physics

Heat convection – five facts

1. Heat convection is heat transfer in a liquid or a gas due to internal circulation of particles.

2. Heat convection occurs only in fluids (liquids or gases).

3. Natural convection occurs because hot fluids, being less dense than cold fluids, rise, whereas cold fluids sink. When a fluid is heated, circulation is caused by hot fluid rising and cold fluid sinking.

4. Forced convection happens when a cold fluid is pumped over a hot surface and takes away energy from the surface.

5. Cooling fins on an engine increase the surface area of the engine and therefore enable more heat convection to occur.

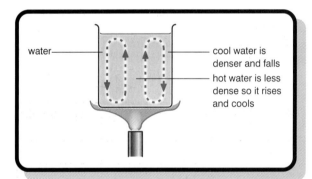

Heat convection in water

Heat radiation – five facts

1. Heat radiation is electromagnetic radiation emitted by any surface at a temperature greater than absolute zero.

2. The hotter a surface is, the more heat radiation it emits.

3. Heat radiation can pass through a vacuum and does not need a substance to carry it.

4. A black surface is a better emitter and absorber of heat radiation than a white surface.

5. A matt (rough) surface is a better emitter and absorber of heat radiation than a shiny surface.

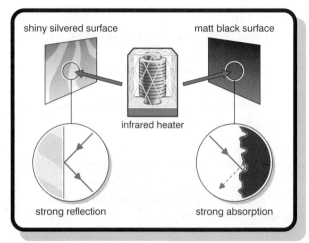

Heat radiation

Heat loss in the home

The rate of heat loss depends on the temperature difference between the inside and the outside of the house. For example, if the temperature inside a house is 25 °C, then heat loss from the house is greater if the outside temperature is 10 °C than if the outside temperature is 20 °C. In the first case there is a 15 °C temperature difference whereas in the second case there is only a 5 °C temperature difference.

Reducing heat loss in the home

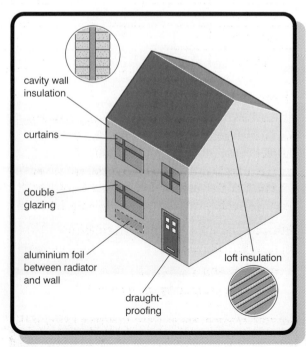

Reducing home heating bills

Energy matters

Conduction
1. To reduce heat loss through floors, fit a thick carpet with underlay.
2. To cut down heat loss through walls, have cavity wall insulation.
3. To reduce heat loss, fit an insulating jacket to the hot water tank.

Convection
1. To reduce heat loss through the ceiling and roof fit thick fibre glass insulation.
2. Fit seals around windows and doors to minimise draughts.

Radiation
1. Cover tanks and pipes with thick insulation.
2. Line walls behind radiators with reflective foil.
3. Make the outer surface of a hot water insulating jacket shiny.

Question

1 Tick boxes in the table to show which heat transfer processes have been reduced by each measure.

	loft insulation	double glazing	cavity wall insulation	radiator foil	heavy curtains	draught excluder
conduction						
convection						
radiation						

How to keep cool

An open-air swimming pool is a good place to be on a hot summer's day. However, standing barefoot on a paved area can be very uncomfortable if the paving material is too hot. It's much cooler in the pool! Why does the water heat up much less than the paving material does?

When a material is heated, the change of temperature depends on

- the mass of the material
- the energy gained by the material
- the **nature** of the material itself
- the physical state of the material (whether it is solid, liquid or gas).

Equal masses of different materials, given the same amount of heat energy, will **not** produce the same rise in temperature. To take account of this fact, materials are assigned a **specific heat capacity** (c) value. This is defined as the amount of heat energy required to raise the temperature of 1 kg of the material by 1 °C.

In symbol form, $c = E_h / m \Delta T$

where c is the specific heat capacity of the material (J/kg °C), m is the mass of material (kg), E_h is the heat energy supplied to the material (J) and ΔT is the temperature rise (or fall) of the material (°C)

Rearranging this equation we can find an expression for heat energy:

$$E_h = cm\Delta T$$

For example, the specific heat capacity of water is 4200 J/kg °C. This means that 1 kg of water must be supplied with 4200 J to raise its temperature by 1 °C. It also means that in order to cool 1 kg of water by 1 °C, 4200 J of energy must be removed from the water.

Answer 1 conduction ✓✓✗✗✗ convection ✗✓✗✗✓✓ radiation ✗✗✗✓✗✗

How not to revise!

Revise Standard Grade Physics

The table below gives the specific heat capacities of some common materials. Note that the specific heat capacity of water is much higher than that of the other materials in this table.

The specific heat capacities of some common substances

material	specific heat capacity J/kg°C
lead	130
copper	380
aluminium	900
oil	2100
water	4200

Worked example 1

A material of mass 3 kg has a specific heat capacity of 500 J/kg°C. How much heat energy is required to raise the temperature from 20°C to 50°C?

Solution

$E_H = cm\Delta T$

$= 500 \times 3 \times 30 = 45\,000$ J

Worked example 2

2 kg of water absorbs 672 kJ of heat in a kettle. If the specific heat capacity of water is 4200 J/kg°C and its initial temperature is 15°C, what will be its final temperature?

Solution

$$E_h = cm\Delta T$$

or $\Delta T = E_h/cm$

$= 672\,000/4\,200 \times 2$

$= 80°C$

Final temperature = 95°C

Question

2 A 50 W low voltage electric immersion heater is to be used to heat 0.20 kg of water in a plastic cup, initially at 5°C.
  a) Calculate the least time taken to heat the water to 45°C.
  b) Give one reason why you might expect it to take longer.

The specific heat capacity of water is 4200 J/kg°C.

Answer
2 a) 672 s b) Heat loss to the surroundings occurs; heat is needed to warm the plastic cup itself.

Measuring the specific heat capacity of a metal block

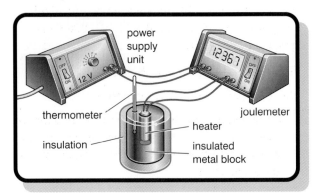

Using a joulemeter

1. The mass m of the block is measured using a top pan balance.
2. The low voltage electric heater and the thermometer are inserted in the block as shown. The block is then insulated.
3. The initial temperature T_1 of the block is measured. The joulemeter reading is recorded.
4. The heater is switched on for 5 minutes
5. The joulemeter reading is recorded again and the highest temperature T_2 reached by the block is measured.

Sample results:

Mass of block = 1.0 kg

Initial temperature = 15°C (T_1)

Initial joulemeter reading = 500 J

Final temperature = 31°C (T_2)

Final joulemeter reading = 15 200 J

Calculation:

Energy E gained by the block = 15 200 – 500

$= 14\,700$ J

Temperature rise $(T_2 - T_1) = 31 - 15 = 16°C$

Rearranging $E = mc(T_2 - T_1)$ gives

$$c = \frac{E}{m(T_2 - T_1)} = \frac{14\,700}{1 \times 16}$$

$= 920$ J/kg°C

114

Energy matters

Change of state

There are three states of matter: solid, liquid and gas. A substance will change state at a particular temperature and when it does this it takes in or gives out heat energy.

Solid → liquid → gas (heat energy taken in)

Gas → liquid → solid (heat energy given out)

★ **When a substance is heated and its temperature rises**, its atoms gain kinetic energy. In a solid, the atoms vibrate more; in a liquid or a gas, the atoms move about faster.

★ **When a substance is heated and its physical state changes**, the potential energy of its atoms changes. When a solid melts, energy is required to let the atoms break free from each other. In a liquid, the atoms move about at random, with weak forces of attraction acting between them. When a liquid vaporises, the atoms or molecules break away from each other completely and only come into contact with each other when they collide

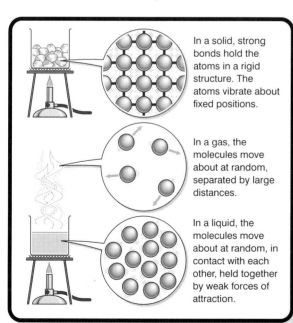

States of matter

a) Use the graph to estimate the melting point of the substance.
b) Describe how the arrangement of the atoms of the substance changed during the cooling process.

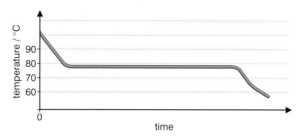

Latent heat

Ice changes from solid to liquid at 0 °C. The heat energy required to do this is called **latent** heat and is taken in to separate the molecules of ice and hence change the state. The temperature remains constant at 0 °C until all the ice melts, i.e. there is no temperature change.

Latent heat of fusion (l_f)

This is the heat energy added to or removed from a material to change it from solid to liquid or liquid to solid at the same temperature.

Latent heat of vaporisation (l_v)

This is the heat energy added to or removed from a material to change it from liquid to gas or gas to liquid at the same temperature.

A refrigerator keeps cool by allowing a liquid to absorb latent heat from the ice box, turning it into a gas.

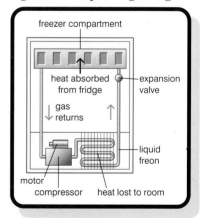

refrigeration process

Question

3 The graph shows how the temperature of a hot substance, initially in the liquid state, changed as it cooled.

Answer 3 a) 78 °C b) Above 78 °C, the atoms move about at random in contact with each other; at 78 °C, they take up fixed positions into which they are locked below this temperature.

The resulting gas is then compressed to turn it back into a liquid. It now gives out heat and the metal fins at the rear of the refrigerator radiate this heat into the room. The process is continually repeated, keeping the inside of the refrigerator cool.

Picnic box coolers use frozen packs of chemicals placed on top of the food to maintain a low temperature inside the box. As the frozen chemicals melt, i.e. go from solid to liquid, they absorb latent heat energy from inside the food and the box, keeping the food cool for a long period of time.

Evaporation

Four facts about evaporation

- ★ Evaporation is the process of a liquid vaporising below its boiling point.
- ★ A hot liquid in an open container cools mainly by evaporation.
- ★ Wet clothing on a washing line dries as water evaporates from it.
- ★ Cooling by evaporation involves transfer of mass as well as transfer of energy from the liquid state.

The process of evaporation

4 **The more energetic molecules can escape from the surface of the liquid.** A molecule at the surface of the liquid must do work to overcome the attraction between it and the other molecules there. Only fast-moving molecules have sufficient kinetic energy to break free of these forces of attraction.

3 **At the surface**, molecules are held in the liquid by the attraction of the other liquid molecules.

1 **In a liquid**, the molecules move about at random.

2 **Weak attractive forces** act between the molecules to keep them together.

Cooling by evaporation

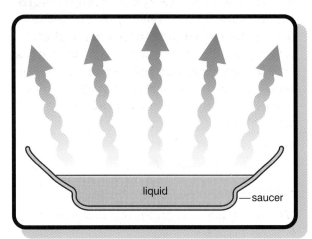

If the vapour formed above a liquid is carried away by a draught of air, the molecules in the vapour state are unable to return to the liquid. The faster-moving molecules in the liquid continuously leave the liquid. Without any supply of heat, the average kinetic energy of the remaining liquid molecules becomes lower and the liquid becomes colder. This process is called **cooling by evaporation**. Two examples of this process are

- the chill factor, experienced by someone outdoors in wet clothing on a windy day
- dabbing a volatile liquid on the skin to numb it before an inoculation.

Conservation of energy principle

This states that energy cannot be created or destroyed and can only be changed from one form into another. Many energy changes or transformations result in the production of heat and hence involve a temperature change.

Electrical energy → heat energy (in a heater)

$E_e = ItV$ $E_h = cm\Delta T$

Kinetic energy → heat energy (in car brakes)

$E_k = \frac{1}{2} mv^2$ $E_h = cm\Delta T$

Potential energy → heat energy (in falling objects)

$E_p = mgh$ $E_h = cm\Delta T$

Work done → heat energy (movement with friction)

$E_w = Fd$ $E_h = cm\Delta T$

Worked example

A motorbike of mass 400 kg is travelling at 20 m/s towards a set of traffic lights. The traffic lights change to red so the brakes are applied, bringing the motorbike to a stop. Assuming all the kinetic energy is changed into heat energy in the brakes, which have a mass of 5 kg, calculate the temperature rise in the brakes (c for brakes material = 400 J/kg °C).

Solution

$E_k = \frac{1}{2} mv^2 = \frac{1}{2} \times 400 \times 20^2$

$= 80\,000$ J

If all the kinetic energy goes into heat energy then $E_h = 80\,000$ J

$E_h = cm\Delta T$

$80\,000 = 400 \times 5 \times \Delta T$

$\Delta T = 80\,000/(400 \times 5)$

$= 40$ °C

Worked example

Oil-filled radiators contain an electrical element that warms up the oil sealed inside the radiators.

a) How much heat energy is required to raise the temperature of 2.5 kg of oil from 20 °C to 45 °C if the specific heat capacity of the oil is 2500 J/kg °C?

b) If the element uses 230 V and takes 12 minutes to heat the oil as described, what is the current flowing in the element?

Solution

a) $E_h = cm\Delta T$

   $= 2500 \times 2.5 \times 25$

   $= 156\,250$ J

b) Assuming all the electrical energy is converted into heat,

$E_e = 156\,250 = ItV$

$I = 156\,250/(12 \times 60 \times 230)$

$I = 0.94$ A

Question

4 A copper water tank of mass of 40 kg contains 125 kg of water at 10 °C. It is heated to 50 °C by means of a 6.0 kW electric immersion heater.

a) Calculate the heat energy needed to raise the temperature of **(i)** the water **(ii)** the copper.

b) Calculate the least time taken to raise the temperature of the water and the tank to 50 °C.

The specific heat capacity of copper is 380 J/kg °C and the specific heat capacity of water is 4200 J/kg °C.

Specific latent heat

Another way not to revise!

An ice lolly needs to be eaten quickly on a hot summer's day. The energy needed to melt 1 g of ice is 340 J, the same as a 100 W light bulb uses in just 3.4 seconds!

The specific latent heat is defined as the heat energy required to change the state of 1 kg of material, i.e.

$l = E_h/m$

where l is the specific latent heat (J/kg), E_H is the heat energy (J) and m is the mass of material (kg)

Rearranging gives

$$E_h = ml$$

Answer

4 a) (i) 21.0 MJ (ii) 0.6 MJ b) 3600 s (= 1 hour)

Revise Standard Grade Physics

★ The specific latent heat of fusion of a material is the energy needed to melt 1 kg of material. For example, the specific latent heat of fusion of ice is 3.34×10^5 J/kg.

★ The specific latent heat of vaporisation of a material is the energy needed to vaporise 1 kg of material. For example, the specific latent heat of vaporisation of water is 2.26×10^6 J/kg.

If the change of state is from solid to liquid or vice versa we must use the specific latent heat of fusion, l_f.

If the change of state is from liquid to gas or vice versa we must use the specific latent heat of vaporisation, l_v.

Worked example
600 g of water is placed in the freezer compartment of a fridge. Calculate the heat energy which must be removed from the water to change it all into ice (specific latent heat of fusion of water = 3.34×10^5 J/kg).

Solution
$E_H = ml_f$
$= 0.6 \times 3.34 \times 10^5$
$= 2 \times 10^5$ J

Worked example
Water in a kettle is brought to the boil at 100 °C and then heated for a further 2 minutes. If the power of the kettle is 1.5 kW, **a)** how much energy is added to the water during the 2 minutes **b)** how much water is changed into steam?

(Specific latent heat of vaporisation of water is 2.26×10^6 J/kg.)

Solution
a) $P = E/t$
$E = P \times t$
$= 1.5 \times 1000 \times 2 \times 60$
$= 180\,000$ J

b) Assuming all the electrical energy goes into heat,
$180\,000 = ml_v$
$m = 180\,000 / 2.26 \times 10^6$
$= 0.08$ kg

Question

5 A 2.5 kW electric kettle contains 3.0 kg of water. The specific latent heat of water is 2.3 MJ/kg. Calculate **a)** the energy needed to boil away 1.2 kg of water at 100 °C **b)** the time taken to supply the necessary energy.

Answer 5 a) 2.8 MJ b) 1120 s

Equations review

1 $E = Pt$ This equation is used in the Transport and Using Electricity sections as well as in this section

Remember that 1 W = 1 J/s

2 % efficiency = useful E_{output}/total $E_{input} \times 100$
or useful P_{output}/total $P_{input} \times 100$

In power stations 'wasted' energy is usually in the form of heat

3 $n_p/n_s = V_p/V_s = I_s/I_p$

This is the 'transformer' equation for 100% efficient transformers

4 $P = I^2R$ This equation is used for calculating the power loss in overhead cables

5 $E_h = cm\Delta T$ c is the specific heat capacity of a substance measured in J/kg °C and ΔT is the difference between the initial and final temperatures

6 $E_h = ml$ l is the specific latent heat of a substance measured in J/kg. If the change in state is from solid to liquid or vice versa, then we use l_f, which is the latent heat of fusion. If the change of state is from liquid to gas or vice versa, then we use l_v, which is the latent heat of vaporisation

round-up

1 a) Which of the following energy sources are renewable and which are non-renewable?
coal   hydroelectricity   uranium   solar   tidal
[5]

b) A kilogram of uranium releases much more energy than a kilogram of coal. State one disadvantage of uranium as an energy source. [1]

2 Home heating bills may be reduced by installing
(i) double glazed windows instead of ordinary windows
(ii) felt insulation in the loft
(iii) draught excluders round the door frames.

a) Which of the above measures reduces heat loss due to conduction and radiation? [2]
b) Which of the above measures reduces heat loss due to convection? [1]

3 A falling weight is used to turn an electricity generator which is used to light a bulb, as shown in the diagram. Complete the energy flow diagram. [5]

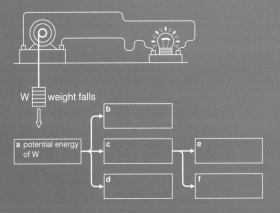

4 An elevator is used in a factory to lift packages each weighing 200 N through a height of 4.0 m from the production line to a loading platform. The elevator is designed to deliver three packages per minute to the loading bay.

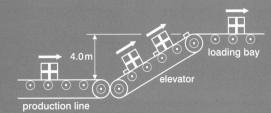

a) Calculate
(i) the potential energy gain of each package [1]
(ii) the work done per second by the elevator to lift the packages. [1]

b) A 200 W electric motor is used to drive the elevator. Calculate the efficiency of the elevator and motor system. [1]

5 A transformer has 1200 turns in its primary coil and 120 turns in its secondary coil.
a) Is this a step-up or a step-down transformer? [1]
b) The primary coil of a transformer is connected to 230 V mains. Calculate the voltage in the secondary coil. [2]

6 A transformer has 60 turns in its primary coil and 1200 turns in its secondary coil. A 240 V, 100 W lamp is connected to the terminals of its secondary coil. The primary coil is connected to an alternating voltage supply.
a) Calculate the voltage of this supply if the lamp lights normally. [1]
b) Calculate the maximum possible current through the primary coil when the 100 W lamp is on. [1]

7 a) State two ways in which the voltage from an alternating current generator would change if its rate of rotation was reduced. [2]
b) Why is alternating current used to transmit electric power through the grid system? [3]

8 a) The diagram shows the construction of a step-down transformer.
 (i) Explain the operation of this transformer. [2]

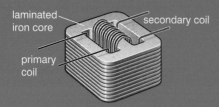

 (ii) Why are the windings of the seconding coil thicker than the windings of the primary coil? [1]
 (iii) Why is the core constructed from laminated iron plates? [2]
b) A step-down transformer has a primary coil with 1200 turns and a secondary coil with 60 turns.
 (i) If the primary coil is connected to 230 V a.c. mains, what will the secondary voltage be? [1]
 (ii) The percentage efficiency of the transformer was measured and found to be 80%. If the primary current is not to exceed 0.1 A, what is the maximum current that can be delivered to a 12 V light bulb connected to the secondary coil? [2]

For questions 9 to 15 use the data in the table on pages 114 and 118 as appropriate.

9 Calculate the energy needed to raise the temperature of
 a) 5.0 kg of water from 15 °C to 50 °C
 b) 5.0 kg of lead from 15 °C to 50 °C
 c) 0.2 kg of aluminium from 10 °C to 60 °C
 d) 0.2 kg of oil from 10 °C to 60 °C. [4]

10 a) An aluminium saucepan of mass 0.15 kg contains 1.2 kg of water at 20 °C. Calculate the energy needed to heat the saucepan and the water to 100 °C.
 b) A copper water tank of mass 12 kg contains 80 kg of water at 22 °C. Calculate the energy needed to heat the water and the tank to 60 °C. [6]

11 A 2.5 kW electric kettle of mass 1.6 kg is made of steel. It contains 0.8 kg of water, initially at 20 °C. The specific heat capacity of steel is 470 J/kg °C. Calculate
 a) the energy needed to heat the kettle and the water to 100 °C
 b) the time it takes to reach 100 °C. [5]

12 Water flows from an electric shower at a rate of 20 g/s. The water enters the shower at 15 °C and leaves it at 45 °C. Calculate
 a) the mass of water leaving the shower in 300 s
 b) the energy needed to heat this mass of water from 15 °C to 42 °C
 c) the power of the electric heater in the shower. [4]

13 The heating element of a 3 kW electric kettle is just immersed when the kettle contains 0.4 kg of water. The kettle is designed to hold 1.6 kg of water when filled. Calculate
 a) the mass of water which would need to boil away after the kettle has been filled in order to expose the heating element
 b) the energy needed to boil away this mass of water
 c) the time taken by the heating element to supply this quantity of energy. [4]

14 Calculate the energy needed to melt 2.5 kg of ice at 0 °C and then raise the temperature of the melted water to 20 °C. [3]

15 In a nuclear power station, water enters the heat exchanger at 20 °C and leaves as steam at 100 °C. Calculate
 a) the energy needed to change 1 kg of water at 20 °C to steam at 100 °C
 b) the mass of water that must enter the heat exchanger each second if heat is to be transferred at a rate of 1000 MJ/s.. [5]

Total = 66 marks

Space physics

7.1 Signals from space

MIND MAP Page 11.

preview

At the end of this topic you will be able to:
- use correctly in context the following terms: 'moon', 'planet', 'sun', 'solar system', 'galaxy' and 'universe'
- state approximate values for the distances from the Earth to the Sun, to the nearest star and to the edge of our galaxy in terms of the time taken for light to cover these distances
- draw a diagram showing the main features of a refracting telescope (objective, eyepiece, light-tight tube)
- state that the objective lens produces an image which is magnified by the eyepiece
- state that the different colours of light correspond to different wavelengths
- list the following colours in order to wavelength: red, green and blue
- state that white light can be split into different colours using a prism
- state that the line spectrum produced by a source provides information about the atoms within the source
- state that there exists a large family of waves with a wide range of wavelengths which all travel at the speed of light
- state that telescopes can be designed to detect radio waves
- use correctly in context the term 'light year'
- draw a ray diagram to show the formation of an image by a magnifying glass
- explain why the brightness of an image depends on the diameter of the objective
- classify as members of the electromagnetic spectrum the following radiations: gamma rays, X-rays, ultraviolet, visible light, infrared, microwaves, TV and radio
- list the above radiations in order of wavelength (and frequency)
- give an example of a detector for each of the above radiations
- explain why different kinds of telescope are used to detect signals from space.

Solar system

The Sun and the nine planets in orbit around it make up the solar system.

Planets in perspective

★ The planets all move round the Sun in the same direction.

★ Their orbits are almost circular, except for Pluto, and in the same plane as the Earth's orbit.

★ The planets reflect sunlight, which is why we can see them.

★ The further a planet is from the Sun, the longer it takes to go round its orbit.

★ When we observe the planets from Earth, they move through the constellations as they go round the Sun.

On another planet

Mercury and **Venus** are called the 'inner planets' as they are closer to the Sun than Earth is. They are rocky, without moons. **Mars** is a rocky planet like the Earth and the inner planets. **Jupiter, Saturn, Uranus** and **Neptune** are giant spinning balls of fluid. **Pluto** is a small rocky planet.

Handy hint

To remember the order of the planets from the Sun, use the mnemonic '**M**y **V**ery **E**asy **M**ethod **J**ust **S**peeds **U**p **N**aming **P**lanets'. Below are some other terms that you should know.

Moon a moon is a body which revolves around a planet.

Planet a planet is a body which revolves around a central star.

Revise Standard Grade Physics

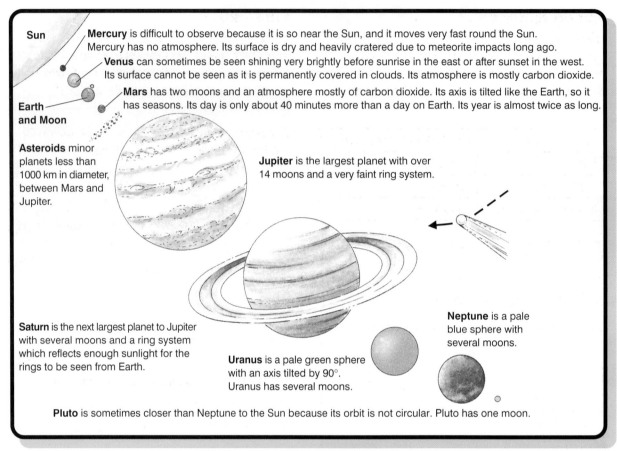

The planets

Sun the Sun is the star at the centre of our solar system.

Star a star is an enormous ball of burning gas which emits light and other radiations into space.

Galaxy a galaxy is composed of around 100 billion stars. The Sun is in the Milky Way galaxy.

Universe the universe is the name given to all the space we know of and can observe. It contains millions of galaxies.

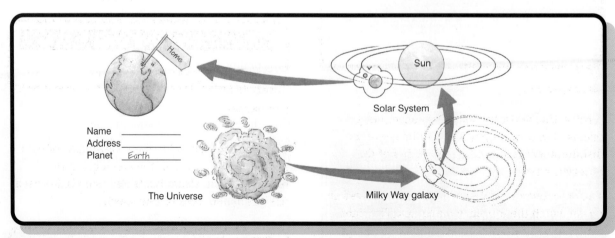

Where do you fit in?

Space distances

Light travels at 300 000 000 (3×10^8) m/s in air or vacuum, but because distances in space are so enormous, light takes various times to reach Earth from stars. Below are three times you should know.

source of light	time for light to reach Earth
Sun	8 minutes
nearest star	4.3 years
far edge of our galaxy	100 000 years

Question

1 Which planet is **a)** nearest to the Sun **b)** the largest?

A refracting telescope

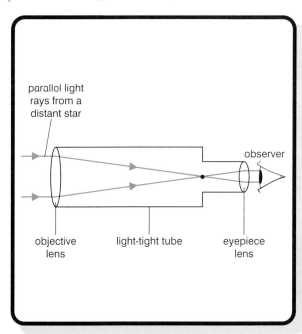

Refracting telescope

A refracting telescope uses two convex lenses enclosed in a light-tight tube. An image of the distant object is formed on the retina of the observer's eye.

The objective lens collects light and produces an image which the eyepiece magnifies. The light-tube can be shortened or lengthened to enable the telescope to be focused.

White light

A triangular glass prism splits white light into different colours. This is because white light is composed of seven different colours of light, each colour having a different wavelength (and therefore frequency).

The prism bends (refracts) long wavelength light least (red) and short wavelength light most (violet).

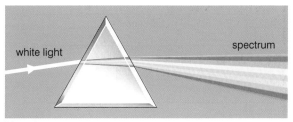

Dispersion

The range of colours produced forms what is called the **visible spectrum**.

Handy hints

★ Remember that **blue bends better than red!**

Red **O**range **Y**ellow **G**reen **B**lue **I**ndigo **V**iolet

Richard **O**f **Y**ork **G**ave **B**attle **I**n **V**ain

Long wavelength ⟶ short wavelength

Low frequency ⟶ high frequency

Line spectra

Some sources of light emit only particular wavelengths of light. When viewed through a spectroscope, they produce a line spectrum.

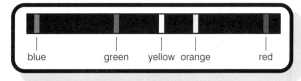

A line spectrum

The lines correspond to the wavelengths of light emitted by the atoms of the elements in the source. Sodium street lights and neon lights used in signs are sources of line spectra.

Answer: 1 a) Mercury b) Jupiter

Each chemical element has its own unique set of lines in its line spectrum pattern. Examination of line spectra can identify different elements and line spectra are used to find the chemical composition of stars.

Stars and galaxies emit long wavelength radio waves and these are detected by large radio telescopes, which are curved dishes with a detector at the focus. The information collected gives information about outer space.

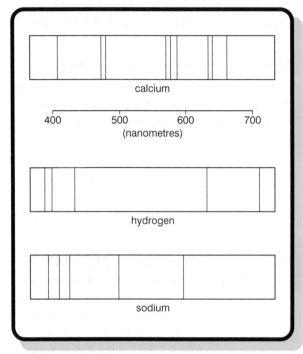

Line spectra of three elements

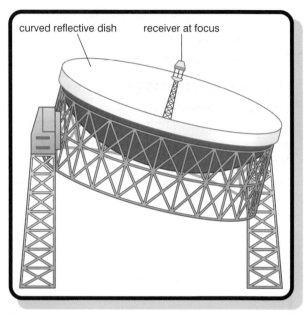

Radio telescope

A large family of waves

Light waves are only a small part of a larger family of waves which all travel at the same speed as light (3×10^8 m/s). The only differences between all the waves forming this larger family of waves are their wavelength and frequency.

Questions

2 a) Does blue light have a longer or a shorter wavelength than red light?
b) Which has the longer wavelength, microwaves or radio waves?

3 a) Name two sources of radio waves from space.
b) Why does the detector in a radio telescope need to be positioned accurately?

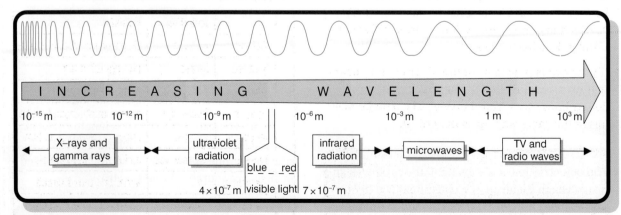

Electromagnetic waves

Space physics

The light year

The distance light travels in 1 year is called a light year. Using the fact that light travels at 3×10^8 m/s and $d = v \times t$, the number of kilometres in a light year can be calculated as follows:

$$d = v \times t = 3 \times 10^8 \times 365 \times 24 \times 60 \times 60$$
$$= 9.5 \times 10^{15} \, \text{m}$$
$$= 9.5 \times 10^{12} \, \text{km}$$

Magnifying an image

The eyepiece lens of a refracting telescope magnifies the image produced by the objective lens. The ray diagram for magnification is shown below.

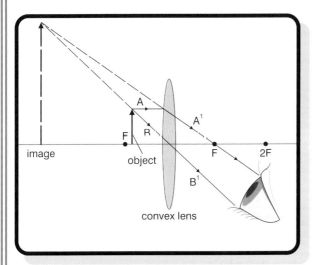

Ray diagram for magnification

Two rays A and B leave the top of the object. Ray B, which passes through the optical centre of the lens, goes straight through to become ray B'. Ray A, which is parallel to the lens axis, is refracted through F to become ray A'.

If your eye is placed as in the diagram, it appears that rays A and B start further back, where the dotted lines cross. You therefore see a magnified image the same way up as the object.

Image brightness

The amount of light reaching a telescope is very small because the stars producing the light are extremely far away. To allow as much light as possible to enter the telescope, the objective lens must be as wide as possible. The bigger the diameter of the objective lens, the brighter the image produced will be.

Detecting signals from space

To provide us with more information about space we have to detect signals produced by distant stars. No single telescope can detect all the different types of radiation being emitted.

A very large radio telescope can detect radio waves emitted by distant stars. It is entirely different from a refracting telescope (which detects visible light) because radio waves have a much longer wavelength than visible light.

The electromagnetic spectrum

The family of waves with different frequencies and wavelengths but travelling at the same speed as light is called the electromagnetic spectrum. These waves are a combination of electric and magnetic waves which travel through a vacuum.

The table below lists all members of the electromagnetic spectrum in order of increasing wavelength (decreasing frequency) along with a suitable detector for each type of radiation. The first column is an easy way of remembering the correct order.

short wavelength

	radiation	detector
Getting	Gamma	Geiger or scintillation counter
X-rayed	X-rays	photographic film
Uncovers	Ultraviolet	fluorescent materials
Various	Visible light	the eye or photographic film
Illnesses	Infrared	special photographic film
Making	Microwaves	aerial and tuned circuit
Treatment	TV	aerial and tuned circuit
Rapid	Radio	aerial and tuned circuit

long wavelength

Answers
2 a) Shorter b) Radio waves
3 a) Stars, galaxies, the Sun b) The dish focuses the radio waves to a point, which is where the detector must be located.

Question

4 a) Why is it possible to see stars with a telescope that cannot be seen without the telescope?

b) State a suitable detector for **(i)** infrared radiation **(ii)** gamma radiation.

Answers

2 a) A star too faint to see directly may be seen using a telescope because the telescope collects more light than the unaided eye **b) (i)** Special photographic film **(ii)** A Geiger tube

7.2 Space travel

preview

At the end of this topic you will be able to:

- state that a rocket is pushed forward because the 'propellant' is pushed back
- explain simple situations involving the rule: A pushes B, B pushes A back
- carry out calculations involving thrust, mass and acceleration
- explain why a rocket motor need not be kept on during interplanetary flight
- state that the force of gravity near the Earth's surface gives all objects the same acceleration (if the effects of air resistance are negligible)
- state that the weight of an object on the Moon or on different planets is different from its weight on Earth
- state that objects in free fall appear weightless
- explain the curved path of a projectile in terms of the force of gravity
- state that an effect of friction is the transformation of E_k into heat
- state that Newton's Third Law is: 'if A exerts a force on B, B exerts an equal but opposite force on A'
- identify Newton pairs in situations involving several forces
- explain the equivalence of acceleration due to gravity and gravitational field strength
- carry out calculations involving the relationship between mass, weight, acceleration due to gravity and/or gravitational field strength including situations where g is not equal to 10 N/kg
- use correctly in context the following terms: 'mass', 'weight', 'inertia', 'gravitational field strength' and 'acceleration due to gravity'
- state that the weight of a body decreases as its distance from the Earth increases
- explain how projectile motion can be treated as two independent motions and solve problems using this method
- explain satellite motion as an extension of projectile motion
- carry out calculations involving the relationships: $E_h = cm\Delta T$, $E_w = Fd$ and $E_k = \frac{1}{2}mv^2$.

Rockets

On the launch pad, a rocket moves because the rocket motor pushes the exhaust gases down (action force) and the exhaust gases push the rocket up (reaction force).

The hot gases are called the **propellant**. The **thrust** of a rocket is the name given to the upward force it exerts and is measured in newtons (N). The equation $F_{UN} = ma$ is used for rocket problems.

Worked example

A rocket produces a combined thrust of 10 000 N. If its mass is 500 kg, find its acceleration.

Solution

$a = F/m = 10\,000/500$

$= 20 \text{ m/s}^2$

Question

1 Calculate the unbalanced force of a rocket of mass 600 kg that leaves the launch pad at an acceleration of 5 m/s².

During launching, the rocket's engines must be used continuously to overcome the Earth's gravitational pull and air resistance. However, once in deep space, since there is no air and no friction to slow the rocket down, the rocket engines can be switched off and the rocket will continue to move in a straight line at a constant speed. This is true for journeys between planets when the rocket has escaped the pull of the Earth's gravity.

Paired forces

Forces occur in pairs. Simple situations happen on Earth as well as in space where this principle is at work all the time, e.g. if you push a drawing pin into a piece of wood, your thumb pushes on the drawing pin but it pushes back on your thumb with an equal force.

Acceleration due to gravity

All falling objects accelerate towards the centre of the Earth. If we ignore the effects of air resistance then any object dropped near the Earth's surface will always accelerate towards it at 10 m/s^2. This means that if an object falls from rest, it will increase its speed by 10 m/s every second, ignoring air resistance and regardless of its mass. This value of acceleration due to gravity, g, is numerically equal to the gravitational field strength g of the earth, which is 10 N/kg.

Weight

The weight of an object is the force of gravity acting on the object. The force of gravity on Earth is 10 newtons for every kilogram of mass. The force of gravity is different on other planets. Although the mass of an object is the same wherever it is, its weight will be different as the force of gravity is different.

For example, on Earth a 70 kg man has a weight of 700 N (70 × 10) since gravity on Earth has a value of 10 N/kg. On the Moon the same man will still have a mass of 70 kg but his weight will be only 112 N (70 × 1.6) since gravity on the Moon is only 1.6 N/kg.

Answer 3000 N

Weightlessness

An object in free fall appears to be weightless. Astronauts inside an orbiting spacecraft appear to be weightless because they, and the spacecraft, are falling towards the Earth with exactly the same acceleration. True weightlessness can only be obtained if an object is so far away from a planet that the gravitational field strength falls to zero.

Projectiles

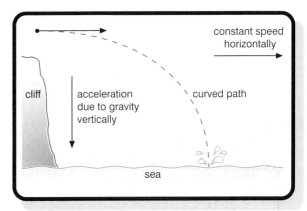

Path of projectile

An object thrown horizontally from a cliff top will follow a curved path towards the sea. The reason for the curved path is that the vertical and horizontal motions combine to produce it. The object will have a **constant horizontal speed** in the absence of friction because there is no force present to slow it down or speed it up. However, as soon as it leaves the cliff top, the force of gravity starts to pull it down and so it will have a **constant vertical acceleration of 10 m/s^2**. The two motions combine to make the path of the projectile a curve, as shown above.

Re-entry of spacecraft

All moving objects have kinetic energy. When a moving object is slowed down, its kinetic energy is converted into other forms of energy, especially heat energy if friction is present.

When a spacecraft returns to Earth from outer space, it is slowed down by friction as it enters the Earth's atmosphere. Its large amount of kinetic energy is converted into great quantities of heat energy.

Question

2 Two students, A and B, each throw a stone horizontally at the same time into the sea from the same place on a cliff top. A's stone hits the sea further away than B's does.
   a) Which stone was thrown fastest?
   b) Which stone hit the sea first?

Newton's Third Law of Motion

This states '**if A exerts a force on B, then B exerts an equal but opposite force on A**'.

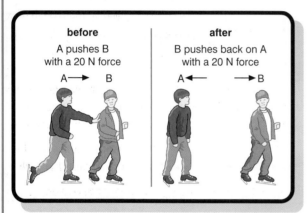

A and B pushing each other on ice

Newton pairs

You are asked to push a car to the side of the road. There are several pairs of forces involved, indicated by dotted circles.

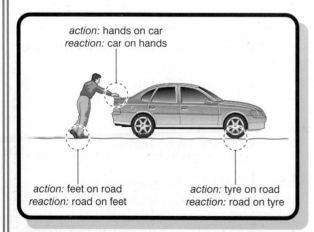

Various action and reaction forces

Answer
2 a) A's b) The stones hit the sea at the same time.

Acceleration due to gravity and gravitational field strength

If we ignore air resistance, all objects near the Earth's surface fall at the same rate regardless of their mass. We call this **acceleration due to gravity** and its value is $10 \, \text{m/s}^2$, which means that the speed of objects increases by $10 \, \text{m/s}$ every second.

The force of gravity on a mass of 1 kg is called the **gravitational field strength** and its value is $10 \, \text{N/kg}$. Note that the values of acceleration due to gravity and gravitational field strength are numerically equal. This can be verified using Newton's Second Law of Motion ($F = ma$) as follows:

Let $m = 1 \, \text{kg}$ and $a = 1 \, \text{m/s}^2$,
then $F = ma = 1 \times 1 = 1 \, \text{N}$

so 1 N will give an object of mass 1 kg an acceleration of $1 \, \text{m/s}^2$

What force, therefore, will give a 1 kg mass an acceleration of $10 \, \text{m/s}^2$?

Answer: $F = ma = 1 \times 10 = 10 \, \text{N}$

i.e. the force required is $10 \, \text{N/kg}$, so if $a = 10 \, \text{m/s}^2$ then $g = 10 \, \text{N/kg}$.

Worked example

a) On Earth $g = 10 \, \text{N/kg}$. What is my weight if my mass on Earth is 65 kg?
b) On the Moon $g = 1.6 \, \text{N/kg}$. What will my weight be on the Moon?
c) What is the value of g on Jupiter if a mass of 100 kg has a weight of 2600 N?
d) What is the value of the acceleration due to gravity on Jupiter?

Solution

a) Weight = mass × gravitational field strength
   $W = mg$
   $= 65 \times 10 = 650 \, \text{N}$
b) $W = mg = 65 \times 1.6 = 104 \, \text{N}$
   *Note: mass (kg) is the amount of matter in a body and **does not** change, i.e. it is always the same.*
c) $g = W/m = 2600/100 = 26 \, \text{N/kg}$
d) The value of acceleration due to gravity on any planet is the same numerically as the value of the gravitational field strength.
   Thus if g on Jupiter is $26 \, \text{N/kg}$ then it is also equal to $26 \, \text{m/s}^2$.

Space physics

> **Question**
>
> **3 a)** Calculate the weight of a 2000 kg space vehicle on the Moon where $g = 1.6$ N/kg.
>
> **b)** Calculate the rocket thrust needed to enable the vehicle to leave the Moon's surface at an initial acceleration of 5 m/s².

Important terms

quantity	symbol	unit	definition
mass	m	kg	The amount of matter in a body
weight	W	N	The force of gravity acting on an object
gravitational field strength	g	N/kg	The force of gravity on a 1 kg mass. On earth $g = 10$ N/kg
acceleration due to gravity	g	m/s²	The rate of change of speed of an object when it falls. On Earth $g = 10$ m/s²

Inertia

The dictionary definition of inertia is **'a state of inactivity or sluggishness'**. In physics, inertia is the tendency of an object to resist any change in motion, i.e. a stationary object is reluctant to move to start with but once it is moving it is then reluctant to stop. The larger the mass, the bigger the inertia. This is why a massive spacecraft keeps going in deep space even when the engines are switched off – its inertia will keep it moving.

Variation of weight with distance from the Earth

The weight of an object **decreases** as it is moved further away from the Earth's surface. This is because the gravitational field strength (g) decreases as the distance from the Earth increases. For example, a 1 kg mass on the Earth's surface has a weight of 10 N, but at a height equal to the diameter of the Earth the same 1 kg mass will have a weight of only 1.1 N (but its mass will still be 1 kg).

Answer 3 a) 3200 N b) 13 200 N

Projectile motion

Satellites which orbit the Earth in a low orbit eventually fall towards the Earth. They are slowed down because of air resistance as they pass through the atmosphere. They follow a projectile path until they finally burn up.

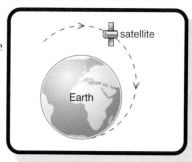

Projectile motion

The motion of any projectile can be treated as **two independent motions**, i.e. the horizontal motion and the vertical motion can be treated separately.

An object in free fall with horizontal motion falls at the same rate as if it had no horizontal motion.

★ Its vertical motion is at constant acceleration, g.

★ Its horizontal motion is at constant speed.

The diagram shows the position at one-second intervals of a ball thrown horizontally from the top of a tall tower. Assuming air resistance is negligible, the ball moves equal distances horizontally each second because gravity does not alter its horizontal motion. However, its vertical motion is the same as if it had been released from rest.

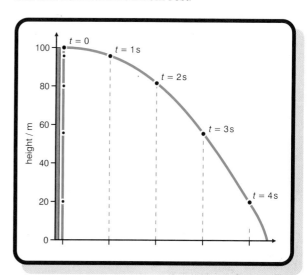

Projectiles

Revise Standard Grade Physics

Worked example

A stone is fired horizontally from a catapult on top of a building at 40 m/s.

a) Show graphically the horizontal speed.
b) Show graphically the vertical speed.
c) Show graphically the combined motion of the stone if it lands on the ground 5 s later.
d) How far away from the bottom of the building does the stone land?
e) What is the height of the building?

Solution

a) Horizontal speed is constant, thus the graph is as shown below.

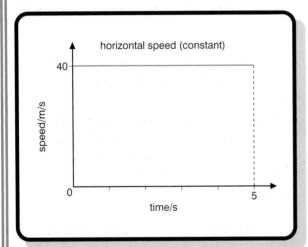

Horizontal speed graph

b) Vertical speed increases by 10 m/s every second.

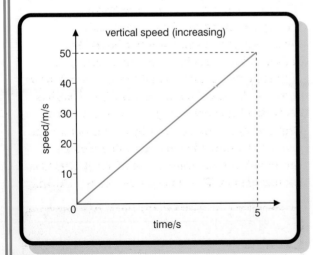

Vertical speed graph

c) Combined motion.

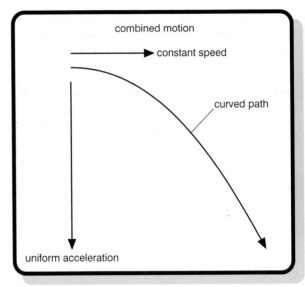

Combined motion graph

d) The horizontal distance the stone travels is the area under the horizontal speed–time graph $= 40 \times 5 = 200$ m. Therefore the stone lands 200 m from the bottom of the building.

e) The vertical distance fallen by the stone is the height of the building, which can be found from the area under the vertical speed–time graph and is given by $\frac{1}{2} \times 50 \times 5 = 125$ m.

Notice how the two motions, the horizontal and the vertical, are treated separately.

Question

4 A ball is thrown horizontally from the top of a tall tower and reaches the ground 30 m away from the foot of the tower 2 s later.

a) Calculate the speed of projection of the ball.
b) Calculate the vertical speed of the ball just before impact.
c) Sketch a graph to show how the vertical speed increases with time.
d) Use your graph to estimate the height of the tower. Assume $g = 10$ m/s².

Answer 4 a) 15 m/s b) 20 m/s c) The graph is a straight line from zero speed at $t = 0$ to 20 m/s at 2 s. d) 20 m

Space physics

Satellite motion

The horizontal launch speed will affect any projectile's path. The larger the horizontal launch speed the greater the horizontal distance travelled.

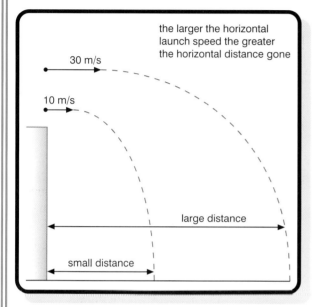

Various launch speeds

However, the distance travelled will be slightly greater than that shown above once the Earth's curvature is taken into account.

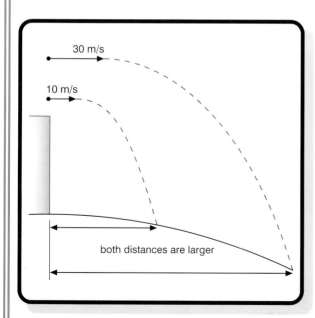

Launch speeds including curvature

At the correct speed, a projectile or satellite can be made to follow the curvature of the Earth, i.e. it will go into orbit.

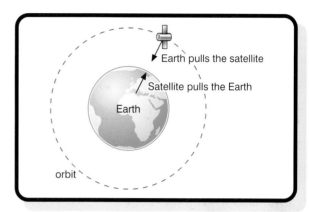

Satellite orbit

If a satellite is in a high orbit outside the Earth's atmosphere, it will not be slowed down and could stay in orbit forever.

Note: At a speed of 8km/s, a satellite falls to the Earth at exactly the same rate as the Earth's curvature falls away, and so the satellite stays at the same height above the Earth.

Question

5 Why does a satellite in a very low orbit return to the ground, whereas a satellite in a much higher orbit does not return?

Re-entry calculations

When a spacecraft re-enters the Earth's atmosphere it is travelling very quickly (approximately 20 000 miles per hour). As it collides with atoms in the air, heat energy is produced due to friction and the spacecraft slows down. In this way kinetic energy ($E_k = \frac{1}{2} mv^2$) is converted into heat energy ($E_h = cm\Delta T$). Some heat energy is carried away by the air but a large amount is absorbed by the heat shield of the spacecraft. A temperature rise of about 1700 °C is not uncommon. The atmosphere exerts a slowing

Answer

5 A satellite in a very low orbit is affected by the Earth's atmosphere, which slows it down and it therefore falls to Earth.

Revise Standard Grade Physics

down **force** on the spacecraft over the **distance** it takes to slow it down. We can therefore involve $E_w = Fd$ in our re-entry calculations.

Worked example
A satellite of mass 60 kg orbits the Earth at a speed of 3 km/s. The satellite is manufactured from a material with a specific heat capacity of 3000 J/kg°C.

(i) Calculate the kinetic energy of the satellite in orbit.
(ii) Calculate the change in temperature of the satellite which might be expected if all its kinetic energy is changed into heat energy as the satellite comes back to Earth.
(iii) If the satellite is slowed down over a distance of 3.8×10^4 m, find the force opposing its motion.

Solution
(i) $E_k = \frac{1}{2} mv^2 = \frac{1}{2} \times 60 \times (3000)^2 = 2.7 \times 10^8$ J
(ii) $E_k = \rightarrow E_h$ i.e. $E_h = 2.7 \times 10^8$ J
From $E_h = cm\Delta T$
$\Delta T = E_h/cm = (2.7 \times 10^8)/(60 \times 3000)$
$= 1500$ °C
(iii) Work done against friction = loss in kinetic energy
$F \times d = 2.7 \times 10^8$
$F \times 3.8 \times 10^4 = 2.7 \times 10^8$
$F = 7.1 \times 10^3$ N

Question
6 A satellite of mass 200 kg moving at a speed of 2000 m/s re-enters the Earth's atmosphere and heats up.
a) What causes the satellite to heat up?
b) What is the kinetic energy of the satellite on re-entry?
c) If all the kinetic energy is converted into heat, calculate the temperature rise of the satellite.

The specific heat capacity of the satellite material is 2500 J/kg°C.

Answer 6 a) friction due to the atmosphere b) 4.0×10^8 J c) 800 °C

The equations used in this section have been met before, for example:

$v = d/t$ and $v = f\lambda$ from Telecommunications

$F = ma$, $W = mg$ and the energy expressions mgh, $\frac{1}{2} mv^2$ and $F \times d$ from Transport, and the energy expression $cm\Delta T$ from Energy.

round-up

Assume $g = 10$ m/s².

1 a) List the following astronomical objects in order of increasing size:
  A Earth **B** galaxy **C** Moon **D** star
  E solar system [5]
b) (i) Which planet in the solar system is furthest from the Sun?
  (ii) Which planet in the solar system is closest to the Sun? [2]

2 The speed of light is 3.0×10^8 m/s.
a) Light takes 8 years to travel from a certain star to the Earth. Calculate the distance from this star to the Earth in **(i)** light years **(ii)** kilometres. [2]
b) Calculate the time it would take in years for a spacecraft travelling at a constant speed of 10 km/s to reach the star in **a)**. [1]

3 a) What is a light year? [1]
b) State to the nearest minute the time taken for light to reach the Earth from the Sun. [1]
c) Where does the energy of a planet come from? [2]
d) Name two forms of energy radiated by a star into space. [2]
e) Light from Uranus takes 2.5 hours to reach us on Earth. How far away is Uranus? [3]
f) Draw a labelled diagram of an optical refracting telescope. [4]

4 a) What is the function of the objective lens of an optical telescope? [2]
b) What is the function of the eyepiece lens of an optical telescope? [2]

5 a) Complete the diagram below to show the path of the two light rays from the tip of an object into the observer's eye. [2]

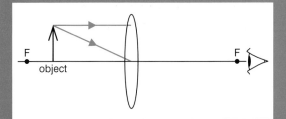

b) Use your diagram to explain why the observer sees a magnified image of the object. [2]

6 a) State two differences and two similarities between an optical telescope and a radio telescope. [4]
b) A student used an optical telescope to observe a group of stars.
(i) Why was the student able to see more stars using the telescope than she could see without the telescope? [2]
(ii) Why did the stars appear to be further apart when observed using the telescope? [2]
(iii) The student used the telescope to observe the Moon. Why was she able to see craters on the lunar surface that she was unable to see without the aid of the telescope? [2]

7 a) State five different colours of the spectrum of white light. [2]
b) Which colour of light, red or blue, has the longer wavelength? [1]

8 a) What is a line spectrum? [2]
b) State one piece of information about a star that can be deduced from its light spectrum. [1]

9 Which two types of electromagnetic radiation most easily reach the Earth's surface? [2]

10 a) How does the weight of an astronaut on board a rocket leaving the Earth change as the rocket moves away from the Earth? [1]
b) A rocket of mass 1200 kg accelerates in space due to a thrust force of 6 000 N from its engines. Calculate the acceleration of the rocket due to this force. [2]

11 Calculate the weight of an astronaut of mass 80 kg
a) on the Earth **b)** on the Moon, where $g = 1.6$ N/kg. [2]

12 An object was projected horizontally at a speed of 15 m/s from the top of a tall tower. The object hit the ground 2 s later.
a) How far did the object travel horizontally from the top of the tower to the ground? [2]
b) What was the vertical speed of the object just before it hit the ground? [2]

13 A stone is thrown horizontally at a speed of 18 m/s from a cliff top and hits the sea 3 s later.
a) (i) What is the horizontal speed of the stone just before it hits the sea?
(ii) What is the vertical speed of the stone just before it hits the sea? [3]
b) What is the horizontal distance between the point of projection of the stone and the point where it falls into the sea? [2]

14 Why does a low altitude satellite fall to Earth after a few weeks or months in orbit, whereas a satellite at much higher altitude remains in orbit indefinitely? [3]

15 A satellite of mass 150 kg moving at a speed of 2000 m/s re-entered the Earth's atmosphere and fell to the ground after travelling a distance of 2000 km through the atmosphere. Calculate
a) the kinetic energy of the satellite before re-entry. [2]
b) its temperature rise, assuming all its kinetic energy was converted to heat and its specific heat capacity was 2500 J/kg °C. [3]

Total = 71 marks

Answers

1 Telecommunications round-up (pages 26–27)

1. a) 3400 Hz (✓)   b) 68 mm (✓)
2. a) 850 m (✓✓)
   b) Sound the siren briefly every 10 seconds and measure the time between each pulse being emitted and its return (✓). The timing becomes shorter if the ship is approaching the cliffs (✓).
3. a) 360/60 (✓) = 60 Hz (✓)
   b) 6 × 30 × 60 (✓) = 10 800 (✓)
4. Wavelength = 40 mm (✓), amplitude = 15 mm (✓)
5. a) Frequency = speed/wavelength = 20 (mm/s)/40 (mm) = 0.50 Hz (✓)   b) 30 (✓)
6. $\lambda = \frac{v}{f} = \frac{1500}{1.5 \times 10^6}$ (✓) = 0.001 m (✓)
7. a) C (✓)   b) B (✓)
8. a) Glass (✓)   b) Greater than (✓)
9. $t = d/v$ (✓) = 100 000/(2 × 10^8) = 0.0005 s (✓)
10. a) 1 5 3 2 4 (✓✓✓✓✓)   b) It makes the audio signal stronger without changing the frequencies (✓).
11. a) One for the video signal (✓) and one for the audio signal (✓)   b) There is one gun for each of the three primary colours (✓).   c) The picture would be in one primary colour (✓).
12. a) 25 × 2 × 60 (✓) = 3000 (✓)   b) 3000 × 625 (✓) = 1.875 × 10^6 (✓)   c) $\frac{1}{25}$ s (✓)
13. Longer wavelengths bend more round a hill compared to shorter wavelengths (✓). Radio waves are longer than TV waves (✓).
14. a) A satellite that orbits the Earth directly above the equator (✓) once every 24 hours (✓).
    b) The metal dish reflects and focuses radio waves (✓) onto the aerial (✓).
15. a) Long wave radio waves follow the Earth's curvature (✓), medium wave radio waves reflect from the ionosphere (✓).
    b) (i) The frequency of electromagnetic waves that carry the audio or TV signals (✓).
       (ii) If they broadcast at the same frequency, the broadcasts interfere with each other where they overlap (✓).
    c) The satellite signals are much weaker (✓) and a dish is used to focus them onto an aerial (✓).

Total = 49 marks

2 Using electricity round-up (pages 48–49)

1. a) (i) Fabric-covered (✓) three-core flex (✓).
      (ii) Plastic-covered (✓) two-core flex (✓).
   b) Live = brown (✓), neutral = blue (✓), earth = green and yellow (✓)
2. a) Electrons go one way only in a d.c. circuit (✓); in an a.c. circuit, the flow of electrons reverses direction every half cycle (✓).
   b) A circuit breaker does not need to be replaced after it has tripped (✓); a fuse needs to be replaced after it has blown (✓).
   c) So that the live wire will be cut off from the appliance if the fuse blows (✓).
3. a) See page 32 (✓✓).
   b) (i) 90C (✓)   (ii) 270J (✓).
4. a)

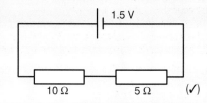

   b) (i) 15 Ω (✓✓)   (ii) 0.1 A (✓) (= 1.5 V/15 Ω)
5. a) 12 Ω (✓✓)
   b) The ammeter reading would double (✓); the voltmeter reading would be unchanged (✓).
6. a) (8 arrangements: ✓ for each two correct arrangements)
   b) Maximum resistance: 2, 3, 6 Ω in series (✓) = 11 Ω (✓); minimum resistance: 2, 3, 6 Ω in parallel (✓) = 1 Ω (✓)

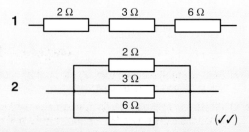

   3, 4, 5 each resistor in series with the other two resistors in parallel.
   6, 7, 8 each resistor in parallel with the other two resistors in series.
7. a) 3 Ω, 0.5 A (✓); 6 Ω, 0.25 A (✓); cell, 0.75 A (✓)
   b) 2 Ω (✓)
8. a) 0.2 A (✓)   b) 1.0 V/0.2 A (✓) = 5.0 Ω (✓)
9. a)

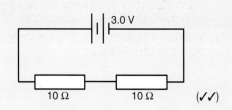

Current = 3.0/(10 + 10) (✓) = 0.15 A (✓);
voltage = 1.5 V for each resistor (✓);
same for both resistors (✓).

b)

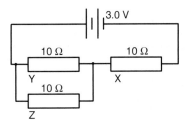

Total circuit resistance = 15 Ω;
cell current = 3.0/15 = 0.2 A (✓)
Current through X = 0.2 A (✓); voltage across X = 2.0 V (✓)
Voltage across Y = voltage across Z = 1 V (✓);
current through Y = current through Z = 0.1 A (✓)

10 a) Electrical energy (✓) is changed to heat (✓) and light (✓).
b) A discharge tube (✓).

11 a) 15.0 V (✓) b) 37.5 W (✓)

12 a) 300 kJ (✓) b) 1000 W/ 230 V (✓) = 4.35 A (✓)

13 a) (i) 1.5 V (✓) (ii) 0.3 A (✓) b) 0.45 W (✓)
c) The current in P becomes less (✓); the current in O becomes greater (✓).

14 a) The same (✓). b) The wire would vibrate (✓) at the same frequency as the alternating current (✓).

Total = 66 marks

3 Health physics round-up (pages 67–69)

1 a) 37 °C (✓) b) (i) Narrower bore (✓) (ii) Constriction or kink in the bore at the lower end (✓).

2 a) (i) Mercury or alcohol (✓) (ii) To produce a magnified image of the thread (✓). (iii) Approximately 32 to 42 °C (✓). b) To detect sounds created in the body (✓) without direct contact between the listener and the patient (✓).

3 a) (i) D (✓) (ii) C (✓) b) Foetal image (✓), image of body organ (✓).

4 a) 0.375 mm (✓✓) b) (i) Without the paste, the ultrasound pulses from the probe would be almost completely reflected at the air-body boundary (✓). With the paste, the pulses from the probe travel directly into the body from the paste without passing into the air (✓). (ii) Reflection at each internal boundary is partial (✓); the pulses spread out as they move away from the probe (✓).

5 a) Towards (✓). b) Less than 30 °C (✓). c) The light ray is totally internally reflected (✓).

6 a)

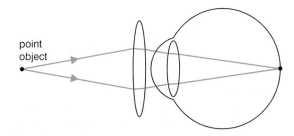

point object

1 mark for each correct ray path (✓✓), 1 mark for showing the rays meeting at the retina (✓).

(b) (i) Convex (✓) (ii) +2.0 (✓) dioptres (or D) (✓)

7 a) Short sight (✓) b) A concave correcting lens is needed (✓), see page 58 (✓✓).

8 a) See page 57 (✓✓)
b) (✓✓)

c) Communications, to carry light pulses (✓); medicine, to see inside the body (✓).

9 a) To detect broken bones or to examine teeth or organs in the body (✓).
b) To produce vitamin D3 in the body or treatment of skin disorders (✓).
c) To detect tumours or poor blood circulation (✓).

10 Alpha, beta and gamma radiation (✓✓✓); alpha radiation is most easily absorbed (✓).

11 Alpha and gamma radiation (✓✓); the tinfoil stops alpha radiation (✓) whereas even thick lead is unable to stop gamma radiation completely (✓).

12 a) (i) See page 64 (✓). (ii) It produces ions (✓). b) X-rays, radioactive waste (✓✓) c) Start the Geiger counter and the stopwatch together. Stop the Geiger counter after exactly 600 seconds (✓). Measure the number of counts recorded by the counter. Divide this number by 600 to give the count rate (✓).

13 Half-life: the time taken for half the atoms of a given radioactive isotope to disintegrate (✓). Isotope: atoms of an element with the same number of neutrons and protons; different isotopes of an element have the same number of protons and different numbers of neutrons (✓).

14 a) (i) Becquerel (✓) (ii) Sievert (or millisievert) (✓) b) (i) 515, 375 (✓) (ii) More than 30 minutes (✓) as the corrected count would take longer to fall to half the initial count (✓) = 0.5 × 515 (✓).

Total = 62 marks

4 Electronics round-up (pages 87–89)

1 Digital: CD player (✓) light-emitting diode (✓) solenoid (✓); analogue: loudspeaker (✓).

2 a) (✓✓)

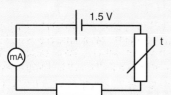

b) The reading would decrease (✓) because the thermistor's resistance would increase (✓).

3 a) (✓)

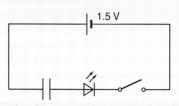

b) Current passes through the LED (making it emit light) as the capacitor charges (✓), and stops when the capacitor is fully charged (✓).

4 a) (i) It decreases (✓). **(ii)** It decreases (✓) **(iii)** It increases (✓).
b) $\frac{500}{1500} \times 9.0\,\text{V}$ (✓) $= 3.0\,\text{V}$ (✓)

5 The LDR's resistance falls when it is illuminated (✓). Therefore its share of the 5 V will fall (✓), causing V_{be} to be > 0.7 V and switch the transistor on (✓), activating the relay, which rings the bell (✓).

6 a) (i) 5.0 − 1.0 V = 4.0 V (✓) **(ii)** 4000 Ω (✓✓✓)
b) The reading would increase (✓) because the LDR resistance would drop (✓), so the current would increase, increasing the voltage across R (✓).

7 a) The resistance of the thermistor increases when it becomes colder (✓). Therefore V_{be} rises to > 0.7 V and the transistor switches on (✓), activating the relay and so switching the heater on.

(b) (i) a diode (✓) **(ii)** electron flow (✓)

8 a) 'OR' (✓)
b)

input 1	input 2	output
0	0	0 (✓)
0	1	1 (✓)
1	0	1 (✓)
1	1	1 (✓)

9 a) The LED lights up when the NOT gate output is 1 (✓). This requires a 0 at the input (✓).
b) To limit the current through the LED (✓), otherwise the LED would fail due to the heating effect of the current through it (✓).

10

input 1	input 2	output
0	0	0 (✓)
0	1	1 (✓)
1	0	0 (✓)
1	1	0 (✓)

11 a)

input A	input B	output
0	0	0 (✓)
0	1	1 (✓)
1	0	0 (✓)
1	1	1 (✓)

b)

input A	input B	output
0	0	0 (✓)
0	1	0 (✓)
1	0	1 (✓)
1	1	0 (✓)

12 a)

window sensor	key sensor	alarm
0	0	0 (✓)
0	1	1 (✓)
1	0	0 (✓)
1	1	0 (✓)

b)

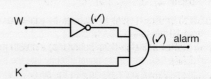

13 a) 0 (✓)
b) The voltmeter reading stays low at first (✓). The capacitor charges up through the resistor (✓) until the capacitor voltage is high enough to switch the OR gate output to logic 1 (✓), which makes the voltmeter reading switch from low to high (✓).

14 a) 4.0 V (✓) **b)** Output current = 4.0 V/10 Ω = 0.40 A (✓), therefore output power = 0.40 × 4.0 = 1.60 W (✓).

15 a) Current = 4.0 V/8.0 Ω (✓✓) = 0.5 A (✓)
b) Power supplied = 0.5 A × 4.0 V (✓) = 2.0 W (✓)

16 a) (i) 30 (✓) **(ii)** Loudspeaker current = 3.0 V/6.0 Ω = 0.5 A (✓). output power = 0.5 A × 3.0 V = 1.5 W (✓) (or output power = $3.0^2/6.0$ (✓) = 1.5 W (✓)
b) (i) Input power = 0.05 A × 0.10 V = 0.005 W (✓)
(ii) Power gain = 1.5 W/0.005 W (✓) = 300 (✓)

Total = 76 marks

5 Transport round-up
(pages 99–101)

1 a) 5 km/h (✓) **b)** 1.4 m/s (✓)

2 a) (i) 4320 m (✓) **(ii)** 4.32 km (✓) **b)** 5.4 m/s (✓)

3 a) The area under the line (✓). **b)** B(✓) A (✓)

4 a) 9.0 m (✓) **b)** -6.0 m/s^2 (✓) **c) (i)** 18.75 m (✓) **(ii)** 27.75 m (✓)

5 a) It was steady (✓) and less than the speed of X for most of the journey (✓). Then the speed of Y was greater than that of X (✓) and Y overtook X (✓).
b) (i) 10 m/s (✓) **(ii)** 15 m/s (✓)

6 a) Instantaneous speed is distance travelled per second over a short time interval (✓). Average speed is total distance travelled divided by time taken over a time interval during which the speed changes (✓).
b) (i) 9 km north (✓) then 24 km south (✓)
(iii) 15 km south (✓).

7 0.5 m/s^2 (✓), 500 N (✓)

8 The force of air resistance increases with speed (✓). As the parachutist falls, his/her speed increases until the force of air resistance is equal and opposite to the weight of the parachutist (✓). The speed therefore becomes constant as the overall force becomes zero (✓).

9 The force of air resistance depends on the speed and shape of the vehicle (✓). The speed increases until the force of air resistance is equal to the engine force (✓). Changing the shape to reduce the force of air resistance enables the car to reach a higher top speed (✓).

10 a) 1.6 m/s^2 (✓) 20 m (✓) **b)** 14 000 N (✓)

11 a) Graph to show a straight line with a negative gradient (✓) from 10 m/s to 0 in 50 s (✓).
b) (i) 250 m (✓) **(ii)** -0.2 m/s^2 (✓) **c)** −6000 N (✓)
d) 0.2 (✓)

12 a) The car skids if the force of the brakes on the wheels exceeds the frictional force of the road on the tyre (✓).
b) The maximum frictional force of the road on the tyres is less on a wet road (✓). This force is equal and opposite to the force of the road on the car (✓), hence the maximum braking force for no skidding must be less on a wet road (✓) so the car will take longer to stop from a given speed on a wet road (✓)

c) (i) 22.5 m (✓) **(ii)** -0.5 m/s^2 (✓), 6000 N (✓)
d) 6 s (✓), 45 m (✓)

13 a) created (✓) destroyed (✓) conserved (✓)
b) $E_p = mgh = 60 \times 10 \times 1.8$ (✓) = 1080 J (✓)

14 a) 80 000 N (✓), 80 000 N (✓) **b) (i)** 6000 m (✓)
(ii) 20 MJ (✓) **c)** It is carried away by the air as heat energy (✓).

15 a) (i) 4000 N (✓), 800 J (✓) **(ii)** 8000 J/s (✓) **b)** Potential energy is converted into thermal energy (or heat) (✓) by the braking force acting on the lift cable (✓). This force is necessary to prevent the lift from falling freely (✓).

16 a) 9.5 m/s (✓) **b)** 2.7 kJ (✓) **c)** 4.5 m (✓) **d)** The pole vaulter uses his or her arm muscles to gain height (✓).

17 a) 2.5 s (✓) **b) (i)** 63 J (✓) **(ii)** 63 J (✓) **(iii)** 31 m (✓)

Total = 74 marks

6 Energy matters round-up
(pages 119–120)

1 a) Renewable: hydroelectricity (✓), tidal (✓), solar (✓); non-renewable: coal (✓), uranium (✓)
b) Waste products are highly radioactive (✓) (or must be stored for many years or reactor is highly radioactive).

2 a) (i) and **(ii)** (✓✓).
b) (iii) only (✓)

3 a) Heat due to friction (✓). **b)** Electrical energy (✓).
c) Sound (✓). **d)** Light (✓). **e)** Heat due to electrical resistance (✓).

4 a) (i) 800 J (✓) **(ii)** 40 J/s (✓) **b)** 0.2 (= 20%) (✓)

5 a) Step-down (✓) **b)** (230 × 120)/1200 (✓) = 23 V (✓)

6 a) 12 V (✓) **b)** 8.4 A (✓)

7 a) The peak voltage would decrease (✓) and the time for one cycle would increase (✓).
b) Alternating current can be stepped up to high voltage on the grid system (✓). The same power can then be delivered with smaller currents than would be necessary if the voltage was low (✓). There is less resistance heating in the grid cables and therefore less power is wasted (✓).

8 a) (i) A changing primary current causes a changing magnetic field in the core (✓). A changing magnetic field through the secondary coil causes an induced voltage in the secondary coil (✓).
(ii) The secondary current is greater since the voltage is stepped down (✓).
(iii) The plates are made of iron to make the magnetic field as strong as possible (✓). The plates are laminated to reduce induced currents in the core (✓).
b) (i) 11.5 V (✓) **(ii)** 1.6 A (✓✓) (✓ for answer 2.0 A)

9 a) 7.4×10^5 J (✓) **b)** 2.3×10^4 J (✓) **c)** 90 000 J (✓)
d) 21 000 J (✓)

10 a) 11 kJ for the aluminium (✓); 403 kJ for the water (✓); 414 kJ in total (✓).
b) 170 kJ for the copper (✓); 1280 kJ for the water (✓); 1450 kJ in total (✓).

11 a) 60 kJ for the steel (✓); 270 kJ for the water (✓); 330 kJ in total (✓) **b)** 330 000 J/2500 W (✓) = 130 s (✓)

12 a) 20 × 300 = 6000 g (✓)
b) 6 × 4200 × (42 − 15) = 680 kJ (✓)
c) 680 000/100 (✓) = 6800 W (✓)

13 a) 1.2 kg (✓)
b) 340 000 × 1.2 = 408 000 J (✓)
c) 408 000/3000 (✓) = 136 s (✓)

14 Energy to melt the ice = 2.5 × 340 000 = 850 kJ (✓); energy to heat water = 2.5 × 4200 × 20 = 210 kJ (✓); total energy needed = 960 kJ (✓)

15 a) Energy needed to heat the water to boiling point = 1 × 4200 × (100 − 20) = 336 000 J = 0.34 MJ (✓); energy needed to vaporise the water = 2.3 × 1 = 2.3 MJ (✓); total energy needed = 2.3 + 0.3(4) = 2.6 MJ (✓)

b) 1000/2.6 (✓) = 380 kg/s (✓)

Total = 66 marks

7 Space physics round-up (pages 132–133)

1 a) C (✓) A (✓) D (✓) E (✓) B (✓) **b) (i)** Pluto (✓) **(ii)** Mercury (✓)

2 a) (i) 8 light years (✓) **(ii)** 7.6×10^{13} km (✓)
b) 240 000 years (✓)

3 a) The distance travelled by light in 1 year (✓).
b) 8 minutes (✓). **c)** From the star (✓) it orbits (✓).
d) Heat (✓), light (✓) **e)** $d = vt = 3 \times 10^8$ (✓) × 2.5 × 60 × 60 (✓) = 2.7×10^{12} m (✓)
f) See below (✓✓✓✓)

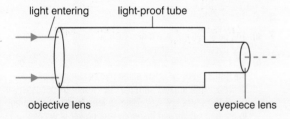

4 a) It forms a real image of the object (✓) in front of the eyepiece (✓).
b) It forms a magnified image of the image formed by the objective (✓) which the observer sees (✓).

5 a) (✓) for each correct ray path (✓✓).

b) The observer sees the tip of the magnified image where the two rays entering the eye appear to come from, as shown by the dotted lines on the diagram above (✓) because the rays appear to come straight from that point (✓).

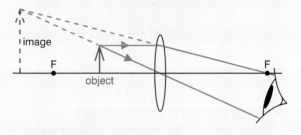

6 a) Similarities: both can be ground based (✓), both detect electromagnetic waves (✓); differences: a radio telescope has an aerial whereas an optical telescope has an eyepiece (✓), radio waves have a much longer wavelength than light waves (✓) (or a radio telescope needs to be much larger than an optical telescope).
b) (i) A telescope is much wider than the eye pupil so it can collect much more light (✓), enabling much fainter objects to be seen (✓).
(ii) A telescope magnifies an extended object (✓) so any two nearby stars seem further apart (✓).
(iii) The telescope gives a magnified view of an object such as a crater (✓), which makes it large enough to see (✓).

7 a) Red, orange, yellow, green, blue, indigo, violet (any 5 ✓✓, 4 ✓), **b)** Red (✓)

8 a) A spectrum of coloured lines (✓) against a black background (✓).
b) Chemical composition or types of atoms or elements present (✓) (or temperature).

9 Light (✓), radio waves (✓).

10 a) It decreases (✓). **b)** a = 6000 N/1200 kg (✓) = 5 m/s² (✓)

11 a) 800 N (✓) **b)** 128 N (✓)

12 a) Horizontal distance = 15 m/s × 2 s (✓) = 30 m (✓)
b) Vertical speed just before impact, $v = u + at$ (✓) = 0 + (10 × 2) = 20 m/s (✓)

13 a) (i) 18 m/s (✓) **(ii)** Vertical speed just before impact, $v = u + at$ (✓) = 10 × 3 = 30 m/s (✓)
b) Distance = 18 m/s × 3 s (✓) = 54 m (✓)

14 The Earth's atmosphere drags on a low-altitude satellite (✓), causing it to slow down and fall to Earth (✓). A high altitude satellite is above the Earth's atmosphere (✓).

15 a) $E_k = 0.5 \times 150 \times 2000^2$ (✓) = 3.0×10^8 J (✓)
b) Use mass × specific heat capacity × temperature rise = energy gained, therefore 150 × 2500 × temperature rise = 3.0×10^8 (✓). Temperature rise = $3.0 \times 10^8/(150 \times 2500)$ (✓) = 800 °C (✓)

Total = 71 marks

Electrical and electronic symbols

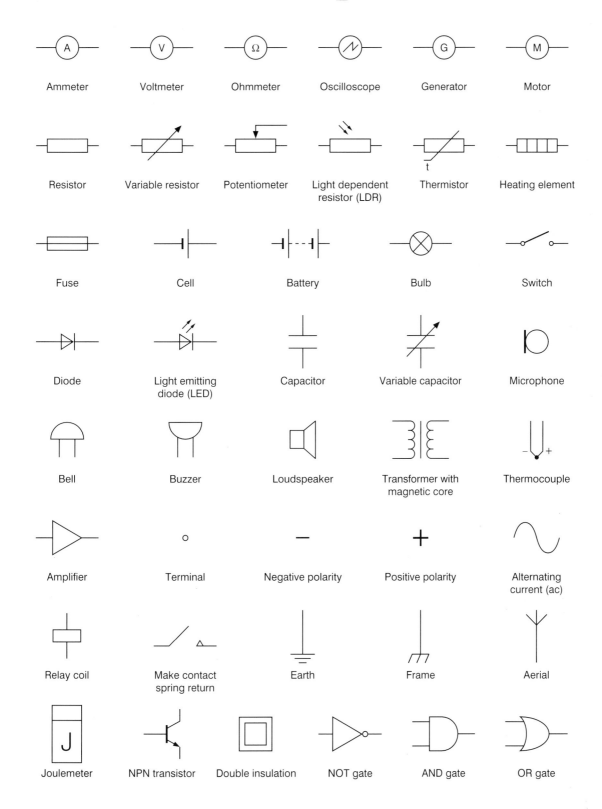

Rearrangement of the formulae

Break for 'Tea' = '<u>T</u>riangle <u>E</u>quation <u>A</u>dvise'

1. Ohm's Law is $V = IR$. The single quantity on the left of the '=' sign goes on the top of the triangle.

2. The other two quantities on the right of the '=' sign go side by side on the bottom of the triangle.

3. To use the triangle, cover up the quantity you wish to use to reveal the correct formula. Always divide the top quantity by the one on the bottom, e.g. $I = \frac{V}{R}$ and $R = \frac{V}{I}$

4. If the quantities left uncovered are on the bottom of the triangle, they should be multiplied, ie. $V = IR$

5. This can be done with the equations like $E_p = mgh$ and $E_k = \frac{1}{2} mv^2$
So, $m = \frac{E_p}{gh}$ and $v^2 = \frac{E_k}{\frac{1}{2}m}$

A different approach is needed for equations of the type $v = \frac{d}{t}$, $P = \frac{E}{t}$, $a = \frac{v-u}{t}$, etc

1. The single quantity on the left of the '=' sign goes to the bottom left of the triangle.

2. Those quantities on the right of the '=' sign go to the top and bottom right of the triangle respectively.

3. So $d = vt$, $v = \frac{d}{t}$ and $t = \frac{d}{v}$.

Physical quantities – their symbols and units

DISTANCE (length)	d	metre (m) – also km, mm, etc
MASS	m	kilogram (kg) – also g, tonne (=1000 kg)
TIME	t	second (s) – also ms, min, h
CURRENT	I	ampere (A) – also mA, etc
TEMPERATURE	T	kelvin (K) – also °C (equal in size to the K)

The above are STANDARD units. Other units are derived from these.

FREQUENCY	f	hertz (Hz) – also 'no of times per s' (= Hz)
SPEED (VELOCITY)	v, u	metre per second (m/s) – also km/h etc
ACCELERATION	a	metre per second squared (m/s²)
FORCE (WEIGHT)	F	newton (N) – also kN, etc
ENERGY (WORK) – potential	E_p	joule (J) – also kJ, MJ, kWh
– kinetic	E_k	
– heat	E_h	
– work	E_w	
POWER	P	watt (W) – also J/s (equal to W), kW, MW
VOLTAGE (POTENTIAL DIFFERENCE)	V	volt (V) – also mV, kV, etc
RESISTANCE	R	ohm (Ω) – also kΩ, MΩ, etc
CHARGE	Q	coulomb (C) – also μC, etc
CAPACITANCE	C	farad (F) – also μF, pF, etc
SPECIFIC HEAT CAPACITY	c	joule per kg K, (J/kgK) – also J/kg °C (same)
LATENT HEAT – fusion (melting)	l_f	joule per kg (J/kg)
– vaporisation (boiling)	l_v	
LENS POWER	P	dioptre (D) – Not the same as POWER above!
RADIO ACTIVITY	A	bequerel (Bq) – also kBq, counts per minute, etc
DOSE EQUIVALENT (absorption of radiation by body)	H	sievert (Sv) – also mSv, μSv

Index

a.c. generator 107, 109
acceleration 91, 92
   due to gravity, g, 127, 128
activity 64
alternating current 32, 34, 43
ammeter 35
amplifier 85
amplitude 13
   modulation 20
analogue devices 70

balanced forces 94
binary numbers 73

capacitor 74, 75
carrier wave 20
CAT scanner 60
change of state 115
charge 33
circuit breaker 44
clinical thermometer 50
clock pulse generator 83, 85
colour mixing 21
copper cables 17
critical angle 18
current 33
curved reflector 23

decibels 52
diffraction 25
digital devices 70
dose equivalent 65
double insulation 30

earth wire 31
electric
   lights 36
   motor 47, 71
electrical
   conductor 33
   flex 29
   insulator 33
   power 36
   safety 30
   symbols 32
electricity costs 43
electromagnetic waves 59, 124, 125
electromagnets 45, 46
endoscope 59
energy changes 96
energy
   conservation of 97
   efficiency 103, 105
   kinetic 98
   nuclear 106
   potential 97
evaporation 116
eye
   image formation 57
   long sight 56
   short sight 56
   sight correction 58
   structure of 54

fibre optics in medicine 57
filament bulb 71
force 93

frequency 13
   bands 23, 25
   modulation 20
friction 93
fuels 102, 106
fuse 29, 43

geiger counter 63, 65
grid system 108

half life 65
heat 96
   conduction 111, 113
   convection 111, 112, 113
   radiation 111, 112, 113
hydroelectricity 104

induced voltage 107
inertia 129
infra-red radiation 60
ionisation 65

lasers in medicine 60
latent heat 115
lens 55
   power 57
light dependent resistor 74
light-emitting diode 71
line spectra 123
logic gate combinations 82, 84
logic gates 82
loudness 16
loudspeaker 71

magnetism 45
mains lighting circuits 43
microphone 73
Morse code 15
motor efect 46

Newton's Third Law of Motion 128
Newton's Laws of Motion 94, 95
noise 52
nucleus, the 62

Ohm's law 37
ohmmeter 40
optical fibre 16
oscilloscope 16

parallel circuits 39
peak voltage 34
pitch 16
planets 121, 122
plug, mains 29
potential difference 38
potentiometer 75
power
   gain 87
   rating 28
   stations 104
projectiles 127, 129
pulse counter 84
pumped storage 104

radio
   receiver 19
   transmission 20, 23
radioactivity
   background 64

   discovery of 61
   properties of 62
   safety 64
   uses 66, 67
reflection of light 16
refraction of light 54, 57
relay 71
renewable energy resouces 102
resistance 35, 38
   heating 37
resistors in combination 41
retina 54, 56
ring circuit 43, 44
rockets 126

satellite
   communications 24
   motion 131
series circuits 39
seven-segment display 72
solar cell 74
solar system 121, 122
solenoid 71
sound 51
specific heat capacity 113, 114
specific latent heat 117
speed of sound 12
speed, average 90, 92
   instantaneous 90, 92
   of a wave 14
   of light 12
speed–time graphs 91, 92
stethoscope 52
switch as an input sensor 79

telephone 15
telescope
   eyepiece 125
   refracting 123, 125
Television
   picture 21
   receiver 19
   transmission 21
   tube 20
thermistor 74
thermocouple 73
thermometer 50
timing circuits 78
total internal reflection 17, 18
transformer 108, 109
transistor 80
transistors in use 81, 84
transmission of electricity 108, 110
truth tables 82, 84

ultra-violet radiation 60
ultrasound 52, 53

variable resistor 35
voltage 34
   divider 75, 76, 77
   gain 86
voltmeter 35

wavelength 13
weight 93, 95, 127
weightlessness 127
work 96, 97

X-rays in medicine 60

Notes

Notes